A Primer in Biological Data Analysis and Visualization Using **R**

A Primer in Biological Data Analysis and Visualization Using R

Second Edition

Gregg Hartvigsen

Columbia University Press   *New York*

Columbia University Press
Publishers Since 1893
New York   Chichester, West Sussex
cup.columbia.edu
Copyright © 2021 Gregg Hartvigsen
All rights reserved

Library of Congress Cataloging-in-Publication Data
Names: Hartvigsen, Gregg, author.
Title: A primer in biological data analysis and visualization using R /
Gregg Hartvigsen.
Description: Second edition. | New York City : Columbia University Press,
2021. | Includes bibliographical references and index.
Identifiers: LCCN 2020053895 (print) | LCCN 2020053896 (ebook) | ISBN
9780231202121 (hardback) | ISBN 9780231202138 (trade paperback) | ISBN
9780231554404 (ebook)
Subjects: LCSH: R (Computer program language) | Mathematical
statistics–Data processing.
Classification: LCC QA276.45.R3 H37 2014 (print) | LCC QA276.45.R3
(ebook) | DDC 005.13/3–dc23
LC record available at https://lccn.loc.gov/2020053895
LC ebook record available at https://lccn.loc.gov/2020053896

Cover design: Milenda Nan Ok Lee
Cover photo: Gregg Hartvigsen

Contents

APPENDIX

Preface to the Second Edition

THIS SECOND EDITION IS AIMED at accomplishing two important tasks. First, I have worked to make necessary corrections, such as a table heading here and a dropped word there. Second, I have received a variety of helpful comments on how to improve the instruction throughout the text. Hopefully, I have succeeded at completing both tasks. Where the book falls short, the blame rests with me.

In this edition I also have worked to improve the chapter-ending questions and have provided solutions to a selection of the problems from each chapter. My goal is to help improve learning and to help faculty who seek to adequately assess student learning through assigning the problems. Many of the solutions include numerical answers so students know they're on the right track. The questions can still be assigned for homework because either the code or answers should be supplied by the students. Below are potential assignments for each chapter:

Chapter	Questions
1	2, 4, 6, 7, 9, 10
2	1a–d, 2b–f, 3
3	1b–e, 2a–d
4	all
5	1, 2, 3, 5
6	1a–c, 2b–d, 3, 4, 5
7	2, 3b–e, 4, 5
8	1a–b, 1e–f, 2a–f
9	1a–e, 2, 3, 4b–d
10	1b–d, 2, 3
11	2, 3b, 4b, 5
12	1, 3, 4, 5, 6, 7

In addition, I have worked to add information that helps readers better understand the tests and techniques implemented in the text. For instance, I've found students curious about why an analysis of variance, which compares different samples, isn't called an "analysis of means." I also have worked to clarify how tests work, particularly the different forms of the apparently simple t-test. The concept of a one- versus two-tailed test is introduced more carefully so readers can both appreciate and implement this approach correctly.

This edition is about 10% shorter than the first edition. It's leaner and meaner. No, I mean it's more concise! It's always tempting to expand the next edition to cover more chapters. I have avoided this urge. New statistical techniques and packages for **R** come out daily, but what early undergraduate biology students need is a short, clear introduction on how to implement and interpret the quantitative methods they're most likely to need. I hope this book meets those needs. Simple methods, such as making graphs using the `plot()` function, are perfectly adequate to produce most publication-quality analyses and visualizations. If and when you need a highly specialized visualization technique, you will, with the basic skills introduced in this text, be able to understand how to implement those procedures.

Acknowledgments

I WOULD LIKE TO THANK the developers and maintainers of the open-source software that was used exclusively to develop this book. I wrote this book in a LaTeX (an open-source typesetting environment) with embedded **R** code (R Core Team, 2020). The original document is in a "noweb" format (`*.Rnw`) that I edited from within RStudio and compiled using the `Sweave()` function. With a single keystroke combination in RStudio, the **R** code is run and written to a LaTeX file and compiled to create the final pdf. Some of the **R** code relies on additional "packages" that were written by volunteers within the vast **R** community.

I also am please to thank members of the SUNY Geneseo community for their support, particularly our provost, Carol Long, for the sabbatical I used to complete the first edition of this book. This second edition was completed during a second sabbatical, so my thanks go, also, to Provost Stacey Robertson. I also would like to thank Jenny Apple, Tom Reho, Bob Simon, Rob Feissner, Jarrod LaFountain, Hayley Martin, Patrick Asselin, Nicholas Whittel, Chris Leary, and the anonymous reviewers for their keen eyes and insightful suggestions as to how to improve this book. For the first edition, my thanks go to my editors and their team at Columbia University Press: Patrick Fitzgerald, Bridget Flannery-McCoy, Anne McCoy, designer Lisa Hamm, and Ellie Thorn. And for this second edition, thanks to Miranda Martin, Brian Smith, Marielle Poss, and Milenda Lee at Columbia University Press, and to Ben Kolstad and Marianne L'Abbate at KGL. My thanks to all for their patience, insight, and belief that I could complete this project. I want you to know that I really tried!

I also must thank the many students that have helped me be a better instructor of biostatistics. In particular, I'd like to thank Briana and Angela Kubik, David Morrison, Yannis Dimitroff, and Tom and Phoebe Hartvigsen for helping point out opportunities to improve the first edition. Without their questions and perplexed looks, this book would have been far more frustrating for readers. Despite all this help, there undoubtedly remain shortcomings in this book, to which I enjoy complete ownership.

On a more personal note, I thank Meredith Drake for her belief in me. There simply is no "without her...." So this one's for you!

Introduction

We face danger whenever information growth outpaces our understanding
of how to process it.

—N. Silver, *The Signal and the Noise*, 2012

Brazil removed from public view months of data on its COVID-19 epidemic on Saturday.

—Reuters, June 6, 2020

So, instead of 25 million tests, let's say we did 10 million tests. We'd look like we were doing much better because we'd have far fewer cases.

—President Donald Trump, June 22, 2020

In our effort to understand and predict patterns and processes in biology, we usually develop an idea or, more formally, a conceptual model of how our system works. We generally frame our models as testable hypotheses that we challenge with data. As the science of biology has matured, our questions about how nature works have gotten more sophisticated and complex. Unfortunately, we are not able to simply look at a table of raw data from an experiment and see an answer to an interesting question with any quantitative level of confidence. To accomplish this, we will learn how to use the **R** statistical and programming software package to process these data (that is, to summarize, analyze, and visualize our results). This approach is laid out in chapter 6. We also will go a step further and work to understand what these results mean biologically.

Data, graphs, and statistics, oh my! Isn't the interesting stuff in biology really just the cool, living things all around us? It is—but it's *so much more beautiful* when we understand it. Maybe you plan to be a doctor or a veterinarian. Perhaps you want to study molecular biology or be an ecologist.

All of these professions require you to fully understand information. You'll encounter daily the need to correctly interpret results from different studies and apply the authors' results within your specialty. You'll also continue to hear news and want to understand what those results mean, maybe also for your everyday life.

We often hear about a variety of global challenges that human beings face. As I write this the human population is facing a pandemic caused by the virus SARS-CoV-2. We're constantly being provided graphs of data that capture the number of cases and the number of deaths. The data are vitally important for us to be able to understand and predict how such challenges are unfolding and how we are to meet those challenges. Hiding or ignoring such data is certainly not going to improve the human condition. This book is aimed at helping you to work with and understand data.

You've probably heard about "coral bleaching." Corals are tiny animals that harbor mutualistic, photosynthetic zooxanthellae (algae) that give the corals their bright colors. Bleaching can occur when water is warmer than normal, an increasingly common phenomenon with global warming. In a study from the Indian Ocean (Southern Hemisphere), corals were found to bleach during the warm summer months (December to March; see figure 0.1).

We can clearly see that zooxanthellae density cycles up and down over time. The authors knew to try and fit a function through these data. But

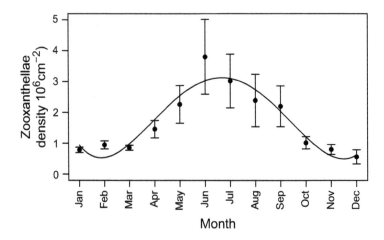

Figure 0.1 The density of mutalistic zooxanthellae in a coral found around the island nation of Mauritius (Indian Ocean). Dips during the summers in the Southern Hemisphere (December to March) coincide with warmer waters that lead corals to expel the zooxanthellae (a.k.a. coral bleaching). Data from Fagoonee et al. (1999).

why do this? As you will learn in this book, we try to fit equations to data like these so that we can complete the two main goals of science:

1. Understand what's going on;
2. Predict what will happen in the future

In this study the authors admit to trying to fit the most complicated function they could through these data (ultimately, a fourth-order polynomial). This line is shown in the graph. This model (the line) does very well at capturing the dynamics of the zooxanthellae density over time but utterly fails to provide us either understanding or any prediction regarding coral bleaching dynamics. The dynamics are undoubtedly cyclical, so the authors should have used a sine curve to identify the average density (1.72 million cm^{-2}) with an amplitude of 1.3 million cm^{-2}, which quantifies the dynamics of coral bleaching in this system. This appropriate model also provides us the ability to predict future (or past) densities as well. A fourth-order polynomial suggests that densities both were and soon will be astronomically high (e.g., > 10 million times their current density in just ten years). My hope is that this book will help you better assess biological information.

What This Book Is (and Isn't)

This book is designed to help you collect, organize, analyze, and visualize data. I assume you have not heard of **R**, and I will, therefore, introduce how to use it to accomplish these goals. Although I imagine you have had some experience making graphs and maybe calculating a few descriptive statistics (e.g., mean and standard deviation on a calulator or in Excel or Google Sheets), I assume you haven't taken a statistics course and have no experience with **R** (we'll install it in the next chapter, in section 1.2).

This book, therefore, aims to give you a foundation upon which to become a better student of science and a better consumer of scientific information. More specifically you will learn how to

- Formulate hypotheses
- Design better experiments
- Do a variety of standard statistical procedures
- Interpret your results
- Create publication-quality visualizations of your results
- Correctly report your results
- Find help so you can become more independent
- Write your own functions and create simple computer programs that will allow you to perform repetitive tasks

Despite being able to do all these things by the end of this short book, it's fair to say that you shouldn't expect to achieve the title of "quantitative guru." Instead, you will work to become competent at finding answers to many of your questions, such as "Are these two samples different?" and "Is there a significant linear relationship between my variables?" You will become a resource to the people around you and be better at understanding and presenting scientific information.

I have written this book in the hope that you'll feel more comfortable with complex biological problems. It has grown out of the challenges facing my own undergraduate students. But it also covers some topics that I think are fun and valuable (e.g., programming). The chapters end with problem sets for you to challenge yourself to use what you have learned. Some of the data are real; some are merely *realistic*. I also have included solutions to many of the problems at the end of the book. Finally, the book is filled with **R** code. You should type this in yourself as you work through the examples because this helps with the learning process. As you'd probably guess, quantitative data analysis isn't a spectator sport! To help you, I also provide all the code at https://github.com/GreggHartvigsen/Primer-Biostats-2e.

This book is neither a formal introduction to **R** nor a statistics textbook, which would be filled with the mathematics of formal statistical theory. Instead, this book helps you to solve problems you're likely to encounter in your undergraduate program in biology. I explain what statistics are and how to interpret and share scientific results in written and visual forms. After working through this book, you should be able to solve a variety of problems with the most widely used statistical and programming environment. I hope you will be less intimidated by data, statistics, and graphs and will be more able to enter data into the computer, test hypotheses, and present your findings.

This book should help you make more appropriate and professional scientific visualizations and discover findings that might have otherwise been missed. You will no longer be satisfied with hearing from anyone things like, "Well, it looks significant" or "There seems to be a trend in the data." For the rest of your career, I hope you become the person who says, "We can test that!"

Who Really Needs This?

In this book, I not only introduce statistical and visualization techniques but also explain why we do all this. There's an unfortunate misconception that we don't really need all this quantitative stuff in biology. I have heard several times the following line of thinking:

Why do we need to use statistics in biology? If the hypothesis is clear, the experiment is designed correctly, and the data are carefully collected, anyone should be able to just look at the data and clearly see whether or not the hypothesis is supported. Statistical procedures are simply safety nets for sloppy science.

As you work your way through this book, you'll see why such thinking limits scientific exploration, understanding, and the ability to make predictions about natural phenomena. Here is a brief list of reasons that statistics, mathematics, and appropriate visualizations are critical for understanding biological systems:

- Statistical procedures allow us to test our ideas rigorously and objectively. We might test whether atmospheric CO_2 concentrations are increasing statistically. We don't address statements of opinion, such as someone stating, "I don't believe global warming is happening." Scientists prefer to assess data rather than opinions.
- Based on our results from data analyses, we often develop formal models that help us to understand and explain how systems work and maybe even make predictions. We want these to be completed correctly, not like the fourth-order polynomial used in the coral bleaching example.
- Biologists often work to understand how multiple factors interact, often in complex ways, to affect biological systems. To determine the individual effects and the combined interactive effects, we need to develop and conduct complex experiments to illuminate biological patterns and mechanisms that cause these patterns. We then use sophisticated data-analysis procedures and visualization techniques to answer these challenging questions.

Biology is one of the more complex sciences. I will admit that, at times, some questions can be pretty simple. Imagine, for instance, that we have 100 randomly selected pea plants and expect a 3:1 phenotypic ratio of yellow to green peas. In this case we should expect to see a ratio of 75 to 25 yellow to green peas. However, we are unlikely to see exactly this ratio. If, instead, we find a ratio of 78:22 we can see immediately (without statistics!) that this is not a 3:1 ratio. Are you prepared, based on this finding, to conclude that this system does not follow the well established rules of segregation? Scientists are predisposed by their profession to be skeptical and, therefore, will not accept a statement like "Trust me that our finding demonstrates conclusively that Mendel was wrong!"

Our goal is to understand biological systems. Unfortunately, anything interesting today is complex (even determining if our data adhere to a simple 3:1 ratio). With quantitative tools, we can better understand how natural systems work. Only then might we be able to make accurate and useful predictions. Science relies on a strong foundation of statistics, mathematics, and the visualization of results, all of which are available to you through the **R** statistical and programming environment.

Additional Resources

There are far too many great sources of information on data analysis, statistics, visualizing information, and programming to list them all here; this book is a basic introduction to all of these topics. I hope you seek more information in all of these areas. If you do, here are a few recommendations that go more deeply into different subsets of the topics covered in this book.

General Introductions to **R**
1. *R: A Language and Environment for Statistical Computing* **R**. R Core Team (**2020**)
2. *An introduction to* **R**. Venables et al. (**2018**)
3. *A beginner's guide to* **R**. Zuur et al. (**2009**)
4. *R for dummies, 2/e*. Meys and de Vries (**2015**)
5. *The R book*. Crawley (**2012**)
6. *R in a nutshell: A desktop quick reference*. Adler (**2012**)

Statistics Books
1. *Biostatistics for the Biological and Health Sciences*. Triola et al. (**2017**)
2. *A primer of ecological statistics*. Gotelli and Ellison (**2012**)
3. *Fundamentals of biostatistics*. Rosner (**2015**)
4. *Biostatistical analysis*. Zar (**2009**)

Programming Using **R**
1. *The art of R programming*. Matloff (**2011**)
2. *Hands-On Programming with R: Write Your Own Functions And Simulations*. Grolemund (**2014**)
3. http://manuals.bioinformatics.ucr.edu/home/programming-in-r

CHAPTER ONE

Introducing Our Software Team

IN SCIENCE, we are interested in understanding systems that are complicated. Our use of quantitative approaches gives us the ability not only to understand these systems but also to predict how a system might behave in the future (or maybe even how it behaved in the past). As we work to understand and predict complex biological systems, we need computational help. You probably have written lab reports using only a calculator. This should be avoided for a variety of important reasons:

- Difficulty in verifying that you entered the data correctly ("I *think* the numbers are right")
- Difficulty in repeating the analysis ("I'm not doing it again because I might get a different answer")
- Inability to share your analytical approaches and results ("Sorry, I hit the all-clear button! You have to trust me")
- Inflexibility in how the data are analyzed ("You wanted me to do what?")
- Inability to make and share appropriate graphs ("Can I take a picture of the graph on my calculator with my phone and incorporate that in my lab report?")

To solve these shortcomings, we will use Excel and **R**.

You may be somewhat familiar with Excel but probably have little or no experience with **R**. Therefore, I welcome you to the world of **R**! I know this might be a scary place for you at first. I bet **R** is really different from all the programs you've used. Fortunately, this introduction is intended for newcomers. But as you proceed, you will learn how to do some really amazing things with **R**. You'll gain independence with practice. **R** is like playing an instrument, participating in a sport, or learning a foreign language—they all require practice. I have confidence that you are capable of using **R** to solve interesting problems. And the more time you spend at it, the better you will get.

1.1   Solving Problems with Excel (or Google Sheets) and **R**

For many analytical problems, we will be able to use **R** by itself. However, in biology, we often test our ideas, or hypotheses, with large amounts of data. Therefore, we will use Excel for what it does well (entering and organizing data). But we will not use Excel to do what it doesn't do well (statistical analyses, modeling, and visualizing data). Instead, these core scientific skills are best done with **R**. If you love Excel, then you'll be happy to know we're not abandoning it—Excel has its place.

It is important to recognize that doing things well is rarely easy. Writing a good poem, playing tennis well, or doing ballet well are all hard. And conducting hypothesis tests correctly and making professional-quality graphs are not simple, one-click operations.

At first you will likely think that making graphs and performing statistical tests in **R** are absolute nightmares. (And when you become a skilled **R** programmer, you'll still be challenged at times!) But the days of skipping an analysis or accepting an ugly or incorrect graph because "that's the best I can do with Excel" are over. You can do it in **R**! Therefore, in this chapter, we will discuss Excel but focus mainly on **R**. The combination of using Excel to organize our data and **R** for analyses and visualizations will allow you to ask and answer questions in biology.

You still may be wondering why you can't just do all of this in Excel. Here is a sampling of reasons why **R** is clearly better than Excel for problem solving in biology. With **R** you can:

- Create professional, publication-quality visualizations.
- conduct quantitative analyses, both analytical and statistical (e.g., do a t-test, solve systems of differential equations, conduct nonlinear regression, use matrix algebra, conduct signal processing, perform wavelet analysis, analyze fMRI data, do genome analyses, and create phylogenetic reconstructions, to name a few).
- Build statistical tests that can be repeated easily and shared with; these tests might rely on their own data, data read from a file, or data acquired directly from a website.
- Do the same thing and work the same way on computers running Mac, Windows, and Linux.
- Write computer programs, such as modeling a population growing over time, using an object-oriented language.

- Access modern analytical tools for biologists that are being developed right now.
- Use and receive widely available help from the **R** open-source community.
- Use open-source software that provides solutions that are "auditable," meaning you can understand and explain to others how you got your results (there are no black boxes; it's open software).
- Write a document like this book; the R environment allows one to compile together in one document words, mathematical equations, computer code, statistical tests and output, and professional-quality graphs, all in the free, open-source LaTeX typesetting environment.
- Carry a research project, paper, all the data, *and* the entire software package for doing the analysis on a low-capacity flash drive.
- Rest assured that your investment in skill building will pay off well into the future; you don't have to hope you'll have access to the program when you move on to your next stage of life.
- Enjoy these benefits because open-source means **R** is free!

Your ability to use **R** to make informed, evidence-based conclusions likely will provide you the most valuable set of skills you'll learn as an undergraduate science major. If you keep this skill set, you will be highly marketable. **R** helps you speak the language of science, which is written in mathematics, statistics, and data evaluation and visualization. This ability to answer scientific questions and present your results professionally is finally in your hands.

Your ability to use **R** helps fulfill an important goal that was synthesized in the report *Scientific Foundations for Future Physicians* produced jointly by the American Association of American Medical Colleges and the Howard Hughes Medical Institute (2009). The authors of this report downplay the importance of memorizing facts and, instead, encourage students to learn to "apply quantitative reasoning and appropriate mathematics to describe or explain phenomena in the natural world."

The report *Vision and Change in Undergraduate Biology: A Call to Action*, produced jointly by the American Association for the Advancement of Science and the National Science Foundation (2009) and still considered an important source for guidelines for undergraduate biology education, identifies six "core competencies" for undergraduates in biology. Below are four of the six competencies that are directly addressed in this book:

1. Ability to apply the process of science (understand how to formulate and test hypotheses).
2. Ability to use quantitative reasoning (e.g., use statistics and quantitative modeling approaches).

3. Ability to use modeling and simulation.
4. Ability to communicate and collaborate across disciplines.

The reason you have this book is to help you achieve these goals. It's time for us to get going.

1.2   Install R and RStudio

We're going to make the installation of your **R** environment a two-phase process. First we will install **R**, which is a basic program with a simple interface. You can do everything discussed in this book in this environment. Consider this the engine, frame, wheels, and steering wheel for a car. It'll get you to wherever you want to go. The second step is to install RStudio, which makes it a much more comfortable ride. For both programs, you can simply accept the defaults offered by the programs during installation.

1. **Install R.** In a web browser, search simply for the letter "r," or go to http://cran.r-project.org/. Follow the instructions to install the correct version of **R** on your computer. Note that if you borrow a computer but you don't have the proper administrative rights, you usually can install **R** on the computer's desktop. If you have a Mac running the OS prior to version 10.6, then the latest version of **R** may not run. Check out the information at http://cran.r-project.org/bin/macosx/ if your installation doesn't work.
2. **Install RStudio.** Back in your web browser, search for "RStudio" or go to http://rstudio.org/. Follow the instructions to install the correct version of RStudio on your computer. Again, the most recent version of RStudio does not seem to work on Mac OS prior to 10.6. For Mac OS 10.5, you can download RStudio at https://s3.amazonaws.com/rstudio-dailybuilds/RStudio-0.95.265. dmg.

Once you have RStudio running, it should look much like the screenshot from a Windows-based computer (figure 1.1). You may see only one large window on the left side. If you do, click on `File -> New` and open a new **R** script file.

RStudio should show four panels. The upper-left panel is where you can enter lines of code into a "script" file. RStudio allows many script files to be open and uses tabs to help you keep track of them. You'll find out more about scripts in section 1.5.

The lower-left panel is the "console" (or "command prompt") where you can type commands and see your answers, like a calculator (see section 1.4).

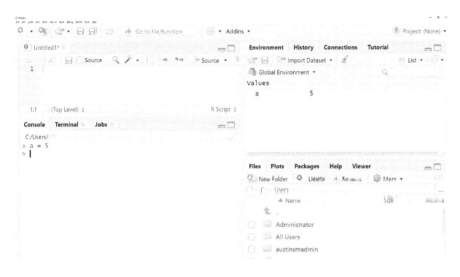

Figure 1.1 Screen shot of RStudio running on the Windows operating system.

Also, if you use a script file and run your commands, the output will appear here in the console.

In the upper-right panel, you'll see variables and their values after you declare them. In the console, assign the number 5 to the variable a with this command, followed by hitting the "Enter" key (←):

```
> a = 5
```

Note that you should not type the ">" character. The a and 5 should show up in the upper-right panel in the "Environment" tab (see figure 1.1). Finally, the lower-right panel displays help information and the plots you create.

1.3   Getting Help with **R**

If you like to get help when using a program, you're in luck. There are many ways to get help with **R** and RStudio. If you don't know how to get the mean of some data or perform an analysis of variance (ANOVA) in **R**, then you can search within RStudio or you might just search on the web for help.

You can get help in the console as follow:

```
> ?mean # Gives help on function mean()
> ??mean # Finds all occurrences of "mean" in the help
> system
> ?"if" # Get help on the control keyword "if" (same for
> "for")
```

Figure 1.2 Adding two numbers on the command line of the console in RStudio.

Do not type the leading ">" character—**R** provides that for you by default in the console. Note that any text that follows the # sign is ignored by **R**. These "comments" can help you, or other readers of your code, to understand what the commands should accomplish.

You also can get help from inside your script files (upper-left panel) by placing the cursor on a keyword and hitting the "Tab" key. A pop-up window should appear with help on your function or similarly spelled functions.

Another important and rich source of help is available online. Feel free to explore this by simply performing a search in a web browser, such as "r mean." For more help you can go to one of these sites:

1. http://cran.r-project.org/doc/manuals/R-intro.html
2. http://www.statmethods.net/
3. https://www.r-bloggers.com/

Or check out the section called "Where Do I Go from Here?" in section 13.1.

1.4   **R** as a Graphing Calculator

Let's begin by running a few commands in RStudio's console. Click inside the console (lower-left panel) to activate it and try adding 2 and 3 (see figure 1.2). Finish by hitting the "Enter" key:

```
> 2 + 3
```

```
[1] 5
```

Did you get 5? You have just run a command at the command prompt in the console. You have used your computer to compute! I hope this has not been painful so far!

The " [1] " before the answer seems a bit strange. **R** is actually reporting that 5 is the first number in a possible array, or vector, of numbers. Sometimes your answer will have lots of values, and **R** will provide you with this counter to help you find values, but more on that later.

Below is a variety of calculations for you to try. The more you practice, the easier this will be. You should check that you get the same answers as I did.

```
> 5-1 # Subtraction
```

```
[1]  4
```

```
> 2*3 # Multiplication
```

```
[1]  6
```

```
> 7/3 # Division
```

```
[1]  2.333333
```

```
> sqrt(9) # Use the sqrt() function to get the square root of 9
```

```
[1]  3
```

```
> 9^2 # 9 squared
```

```
[1]  81
```

```
> log(3) # Natural logarithm of 3
```

```
[1]  1.098612
```

```
> log(3,10) # Log of 3 (base 10), or use log10(3)
```

```
[1]  0.4771213
```

What if a command doesn't work? **R** is really picky about how you enter commands. There's a little wiggle room with spaces, for instance, 5+3 and 5 + 3 both work, but **R** is very finicky about nonspace characters. For example, sqrt{9} doesn't work because braces are different from parentheses. **R** will give you an error message if something's not quite right:

```
> sqrt{9}
Error: unexpected '{' in "sqrt{"
```

Computers, and programs like **R**, generally do exactly what you tell them to do, which might not be what you intended them to do! If something goes wrong, **R** will return an error message that should be somewhat helpful, as you saw above, but it's never very friendly about it. It's important not to take this personally—for **R**, it's all business.

Let's try more complicated calculations. The following lines of code rely on you providing some data in an "array," which is a group of objects that are of the same type, like all numbers or all letters, packaged together. Once the data are in an array, then far more interesting things can happen. Be sure to try the following examples because we'll be using arrays throughout the rest of the book.

```
> 1:5 # create an array of integers from 1 to 5
```

```
[1] 1 2 3 4 5
```

The ":" in the above expression is an operator that makes an integer array ranging from the first value to the second value. Often, however, we have different numbers that we want to combine. We can use the combine function (c()) to group any set of numbers together into an array.

```
> c(1, 2.5, 3, 4, 3.5) # combining five numbers into an array
```

```
[1] 1.0 2.5 3.0 4.0 3.5
```

We can store those numbers in a variable so that we can use them again later (see box 1.1 for more on variables). Here's how to store the numbers into the variable dat:

```
> dat = c(1, 2.5, 3, 4, 3.5) # store numbers the variable
> "dat   "
```

Above is how this book "assigns" information to a variable (or object; see box 1.1). The **R** language also permits the symbol "<-" (a "less than" sign followed by a hyphen) to be used for assignments. The following is an example that is equivalent to the assignment above:

```
> dat <- c(1, 2.5, 3, 4, 3.5) # store numbers the variable
> "dat   "
```

You can use either of these methods. Whichever you choose, however, you should be consistent throughout your work.

Note that when you run this line of code, there's no output provided. That's because **R** has completed the task: store the numbers as an array in the variable dat. If you want to see the contents of an array, such as dat, you can type the variable name in the console and hit <enter>. Try it:

```
> dat
```

```
[1] 1.0 2.5 3.0 4.0 3.5
```

> **Box 1.1. *What is a variable?*** Variables, also called objects in **R**, are letters or words that store information. We usually use them to store a number or a group of numbers for later use, as we just did in this chapter for the data in the variable dat. Alternatively, variables can store characters. For example, we could store Darwin's Origin of Species (OoS) in the variable "OoS":
>
> ```
> > OoS = c("When", "on", "board", "H.M.S.", "Beagle....")
> ```
>
> Variable names should be as short and descriptive as possible (but no shorter). They should always begin with a letter (never a number or symbol). Descriptive names will help your future self and others who might look at your code. Good names and commented code help save us time. If, for example, you have the mass of a dog, don't use the variable "a" but instead a more meaningful name with words separated by periods:
>
> ```
> > dog.mass = 25.2 # a 25.2 kg dog
> ```

Storing numbers in an array variable is very common in statistics. We'll learn more about variables when we discuss data in chapter 2. Now that the data are in the array called "dat," we can perform a variety of operations on them. Try these:

```
> sum(dat) # sum up all values in array "dat"
```

```
[1] 14
```

```
> length(dat) # tells you how many numbers are in "dat"
```

```
[1] 5
```

```
> sum(dat)/length(dat) # this calculates the mean
```

```
[1] 2.8
```

```
> summary(dat) # more descriptive statistics for "dat"
```

```
   Min. 1st Qu.  Median    Mean 3rd Qu.    Max.
    1.0     2.5     3.0     2.8     3.5     4.0
```

```
> dat/5 # divide each value in "dat" by 5
```

```
[1] 0.2 0.5 0.6 0.8 0.7
```

```
> dat[5] # returns the fifth element in array "dat"
```

```
[1] 3.5
```

```
> dat[1:3] # returns the first three elements of array "dat"
```

```
[1] 1.0 2.5 3.0
```

Sometimes you will need to make a longer sequence of numbers. We can do that using the `seq()` function. The following line prints the sequence to the screen.

```
> seq(1,10, by = 0.5) # sequence from 1 to 10 by 0.5
```

```
 [1]  1.0  1.5  2.0  2.5  3.0  3.5  4.0  4.5  5.0  5.5  6.0
[12]  6.5  7.0  7.5  8.0  8.5  9.0  9.5 10.0
```

The next line of code, instead, stores the sequence in a variable called `my.seq`.

```
> my.seq = seq(1,10, by = 0.5) # store result in variable
> "my.seq"
```

We can see that **R** uses square brackets to represent the index number for arrays. From the above output, we see that the twelfth number is 6.5. We can get that number by using this index value in square brackets at the end of the variable's name.

```
> my.seq[12]
```

```
[1] 6.5
```

Note that indexing in **R** begins with `[1]`, which differs from some other programming languages, such as C or `Python`, which start with zero. This conforms with mathematical expressions, such as this: $\sum_{i=1}^{N}$ `my.seq`$_i$ = 104.5.

Sometimes, instead of a sequence of numbers, we need to repeat numbers, or even letters, for different experimental designs. Let's try this using the `rep()` function:

```
> rep(c("A","B","C"), times = 2) # entire array twice
```

```
[1] "A" "B" "C" "A" "B" "C"
```

```
> rep(c("A","B","C"), each = 2) # each element twice
```

```
[1] "A" "A" "B" "B" "C" "C"
```

You also can combine calculations within the declaration of an array.

```
> p = c(1/2,1/4,1/4) # three proportions saved in an array
> p
```

```
[1] 0.50 0.25 0.25
```

R also is good at making graphs. Sometimes you need to see what a sine curve looks like or, perhaps, a simple polynomial. Suppose you are asked (or simply want) to view a function such as this:

$$y = 2x^2 + 4x - 7 \tag{1.1}$$

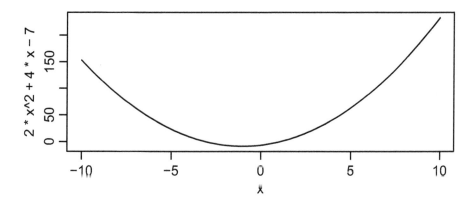

Figure 1.3 The graph of the function $y = 2x^2 + 4x - 7$ over the range $-10 \le x \le 10$, made using the curve() function.

over the range $-10 \le x \le 10$. In **R**, it's really easy! You can use the curve() function (see figure 1.3):

```
> curve(2*x^2 + 4*x - 7,-10,10)
```

The curve() function requires us to provide the right-hand side of the equation, which must include "x." You also might include the range if you want something other than a graph from 0 to 1. We can provide additional arguments if we want to enhance the graph (see box 5.2). If entered correctly, **R** will create a smooth curve over the provided range (figure 1.3). We will use the curve() function to help us solve some tough problems later in the book (e.g., section 11.3).

Here's one more thing to try. Histograms are great graphs, as we'll see, to visualize the distribution of data. The elusive bell-shaped curve of the normal distribution can be made using data drawn from the standard normal distribution (mean = 0 and standard deviation = 1).

Here's how to get 10,000 values from the standard normal distribution and send them to the hist() function:

```
> hist(rnorm(10000))
```

I didn't reproduce the graph because we'll work on this later (see chapter 5). But if you typed that correctly at the command prompt and hit <enter> I hope you saw a cool "histogram." What's happening under the hood is the following. First, rnorm() is a built-in **R** "function." Functions are routines that do a bunch of stuff behind the scenes (see box 1.2). In the rnorm(10000) call above, we send the number 10,000 (without the comma) to the rnorm() function as a single "argument." The function rnorm() then "returns" an array of $10,000$ random numbers, drawn from a "standard normal distribution" (mean = 0, standard deviation = 1). The call above then uses those $10,000$ values as the "argument" to the hist()

function. The `hist()` function then divides the numbers into bins and, behind the scenes, calls the `plot()` function that makes the histogram, displayed in the graphics window. There's a lot going on with only that one, simple line of code!

1.5   Using Script Files

The command line in the console (figure 1.2) is great for quick answers to simple, one-line problems. For most exercises, however, it is best to create and use a "script" file. This is a plain text file, created from within RStudio, that contains a series of commands, each on its own line, that can be run either one command at a time or all together as a "batch" job.

Box 1.2. Functions. In programming languages, like **R**, a function is a defined set of instructions that does something. A function may or may not require "arguments" as input. Multiple arguments need to be separated by commas. Functions may return a variable and/or perform an operation (e.g., make a graph). The `sum()` function, for instance, adds together an array of numbers and returns the total:

```
> sum(1,2,3,4)
```
```
[1] 10
```

R provides many functions that simplify your work. You can even write your own functions, too (see section 11.1).

To run a line of code in a script file (see box 1.3), you need to have the file open in RStudio. Place the cursor anywhere in the line you want to run and hit `<ctrl><enter>` (Windows PC) or `<cmd><enter>` (Mac). You also can highlight any chunk of code, highlighting like you do in other programs, and run only that which you have highlighted. I will often want to run the entire script file and so will hit `<ctrl><a>`, which highlights everything, then `<ctrl><enter>` (replace `<ctrl>` with `<cmd>` on a Mac).

Box 1.3. Create a script file. Click on the following in RStudio: `File → New File → R Script`. An empty script file should appear in the upper-left panel in RStudio (by default). Save it with a meaningful name in a new folder for this exercise (`<ctrl><s>` or `<cmd><s>`). You may need to create that folder, too, which could hold data files, a presentation file, and maybe a lab report or paper. You can see what the work flow is for a typical **R** project, such as a laboratory exercise, in box 1.4.

Note that the **R** code needs to be correct for it to work. Otherwise, **R** will provide an error message in the console. If you forget a comma or spell the sqrt() function SQRT(), **R** will provide you a mildly helpful error message. You also should include "comments" to yourself after the # character. Comments are ignored by **R**. Try the example that you can see in figure 1.4.

Organizing your code into script files and placing those into well-named project folders will *greatly* simplify your life. Be organized from the start—it will save you time. One last beauty of RStudio is that when you close it and later open it up, it will reopen the files that you left open and place the cursor where you left it.

Box 1.4. *What does a typical R project look like?* Let's imagine that you're in a laboratory class and several groups collected and pooled their data. The data have been made available in an Excel spreadsheet in the .CSV file format (one of the options in both Excel and Google Sheets). Here are the steps you might take to complete the lab:

1. Download the data file and store it in a folder for this class/lab/ project.
2. Open RStudio and create a new script file, give it a meaningful name, and save it in the project folder with the data.
3. Set the working directory in RStudio to the location of your script file (Session -> Set Working Directory -> To Source File Location).
4. Write yourself some commented text (lines that start with the # symbol) about the project so when you return you'll know what you were trying to do.
5. Write lines of code that do the following:

   (a) Read in the data from the spreadsheet file (see section 2.2).
   (b) Explore your data. For example, you will likely gather summary statistics (see chapter 4), test whether your data are normally distributed (see section 4.4), and/or create visualizations (see chapter 5).
   (c) Conduct the appropriate statistical test(s) (see chapters 7–11).

6. Once you've completed the tasks above, you'll need to copy your graphs and statistical output from RStudio to your laboratory report.

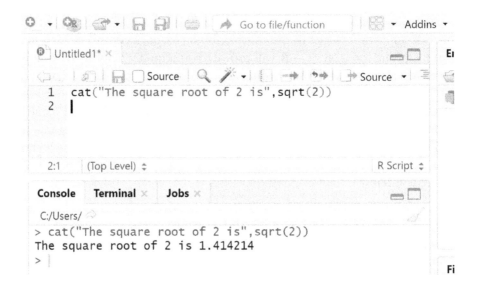

Figure 1.4 RStudio after running the script file that is in the upper-left panel. The output is in the console (lower-left panel).

1.6   Extensibility

When you install **R** and RStudio, you will have the basic version of the software. As you progress in your use of **R**, you'll likely need to add more tools (or get the latest version of your existing tools). To get those tools, you will install "packages" from internet servers located around the world. Once these packages are installed, you have to "load" them with the `library()` command at the beginning of your session.

At first this probably makes **R** seem like an incomplete software package. It's actually a very efficient use of your computer and its resources. You only install what you need. The major computer packages with which you are familiar require a lot of time to install and consume a lot of space on a hard drive. **R** is much smaller, faster, and customized by you for what you do.

When you try to install a package for the first time during an RStudio session, you may be asked to choose a "mirror" website, so you need to be connected to the internet. You should choose one that is geographically nearby (although, at times, I'll go international for fun). Once these packages are installed, you can use them without being connected to the internet. You also can save the packages to a flash drive if you're at an internet cafe in an exotic study abroad location and install them later. Here are the packages I will assume you have installed: `e1071`, `deSolve`, `plotrix`, and `UsingR`. Note that names are case sensitive. To install the `UsingR` package, type:

```
> install.packages("UsingR")
```

and to load it into **R** for your current session, type:

```
> library(UsingR)
```

Note that you need quotes around the package name when you use the `install.packages()` function but not in the `library()` function.

If you are using an older version of **R**, then you will be warned that the package was compiled using a more recent version of **R**. This warning is not likely to cause a problem. If you get this message, you should install the latest version of **R**, at your convenience. Note that "warnings" are just warnings. You are likely fine. An error message, however, means something is wrong and you'll have to fix it.

Now that you've installed **R** and RStudio, created a script file, completed some calculations, and installed a few packages, it's time to test-drive **R**. Below are a few problems for you to try.

1.7   Problems

1. Find the solutions to the following problems:

    (a) $\sqrt{17} = ?$
    (b) $\log_8(10) = ?$ That's the log of 10, base 8. Read the help on the `log()` function (> ?log)
    (c) What's e^{10}, where e is the base of the natural logarithm? If you enter > e^10 in your script file and run it, you'll get an error. Why? Do a quick search on the internet to find how to raise e to a power or, as above, query **R** with "> ?log".
    (d) If $x = 3$, find y, given the following expression.

    $$y = \frac{1}{17} + (5x + 7)^2 + ln(17)$$

    Note that the natural logarithm (ln) of a number is obtained in **R** using the `log()` function.

2. Find the basal area of a tree (cross-sectional area of a tree trunk) if the diameter is 13.5 cm. *Hint:* Area $= \pi r^2$ and that, in **R**, the constant `pi` holds the value of π.

3. Calculate the volume of the Earth in km^3. It is an oblate spheroid, which is a somewhat flattened beach ball. The volume of an oblate spheroid is $V = \frac{4}{3}\pi a^2 b$ where a is the radius to the equator and b is the radius to one of the poles. Here are these values for Earth in kilometers:

```
> a = 6378.137
> b = 6356.752
```

4. The following are masses for chickadee chicks from several nests, measured in grams:

$$3.2, 6.7, 5.5, 3.1, 4.2, 7.3, 6.0, 8.8, 5.8, 4.6$$

   (a) Combine the data into a single variable array called my.dat using the c() function. Use **R** code in an **R** script file to answer these questions.
   (b) What is the total mass of all the chicks combined?
   (c) It's easy enough to count, but use the length() function to count how many chicks are in the sample.
   (d) Without using the mean() function, calculate the mean chick mass in grams. Note you calculated the total mass and the number of chicks above.

5. Create an array of numbers from 0 to 5, by steps of 0.5, using the function seq(). Save the result in a variable called i. Using code in your script file, write the values of 2^i in the console (it should be an array of numbers).

6. A population of seagulls has 47 females and 38 males. What proportion of birds are females?

7. You are riding in your friend's monster truck and you've stopped yet again at the gas station to fill up. You wonder what the gas mileage is for this vehicle. Your friend tells you it went 230 miles on the 28 gallons of gas, the amount just put in the tank.

   (a) What's the fuel efficiency for this vehicle in miles per gallon (mpg)?
   (b) If the tank holds 32 gallons, how far can this vehicle be driven?

8. There are many datasets built into **R** (type data() to see them in a text file within RStudio). You can learn more about each with the help feature. There's an interesting time series dataset on the ring widths measured from an ancient bristlecone pine tree (*Pinus longaeva*). You can learn more about these data with this command:

```
> ?treering
```

Answer the following questions using these data:

   (a) The help on this dataset (> ?treering) tells us how many observations there are. Provide code that confirms or denies this.
   (b) Graph the data by sending these to the plot() function.
   (c) Tree ring widths are generally correlated with climate (e.g., amount of rainfall). The data are bouncing around a lot. Is there a periodic pattern hidden in these data? If so, what's the

dominant period in years? Below is a technique that tweaks out of a time series dataset like this long-term signals of a repeating pattern.

```
> the.spectrum = spectrum(treering) # spectral analysis
> # get the maximum spectrum
> max.spectrum = which(the.spectrum$spec ==
> max(the.spectrum$spec))
> # the period = 1/dominant frequency
> 1/the.spectrum$freq[max.spectrum]
```

What is the dominant period in years?

9. Graph the Michaelis-Menton function for enzyme kinetics. It is represented by this equation:

$$v = \frac{V_{max} \cdot x}{K_m + [x]}$$

where v is the velocity of the reaction, V_{max} is the maximum velocity, x is the concentration of the substrate (usually denoted as $[S]$), and K_m is the Michaelis-Menton constant. Use the curve() function of this relationship for concentrations of $x \in (0, 5)$, assuming that $V_{max} = 0.9$ and $K_m = 0.5$.

In chapter 11 we discuss the analysis of this relationship in greater detail (see section 11.3).

10. A population of koalas in Australia has a per-capita growth rate of $\lambda = 0.95$ per year. They grow according to this equation:

$$N_{2020+t} = \lambda^t N_{2020}$$

There are currently 127 koalas now ($N_{2020} = 127$).

(a) How many koalas will there be in the year 2025 (N_{2025})? In an **R** script file (in RStudio), assign the correct values to variables called lambda, t, and N.2020 (or whatever year it currently is). Each of these should be placed on their own line in your script file. On the fourth line perform the calculation that prints how many koalas there will be in 2025 (N_{2025}).

(b) How many were there in 2015?

CHAPTER TWO

Getting Data into **R**

THERE ARE MANY WAYS TO GET DATA INTO **R**, and which technique you use depends on a variety of factors. All data are eventually stored in "variables" in **R**. A variable is a named object in **R** that is used to store and reference information (see box 1.1). The method we use to do this usually depends on the amount of data we have. The three basic approaches are as follows:

1. Fewer than about 25 values: create a variable and assign data to that variable using the `c()` function.
2. More than about 25 values: enter data into a spreadsheet, save the file in a `.csv` format, and then read the data file into **R** using the `read.csv()` function.
3. Read data in from a website, usually using `read.csv()` (comma separated variables) or `read.table()` for other types of data files.

We'll discuss each of these methods in more detail in the coming sections. Note that getting data into **R** accurately is a critical step in understanding what the data tell us. Extreme care needs to be taken in this step or all our efforts in doing the work could be lost. It's always a good idea to have someone help you with double-checking that the data have been entered into the computer correctly.

2.1 Using `c()` for Small Datasets

The best way to get small amounts of data into **R** is to use the combine function (`c()`) in a script file. This function, as it sounds, groups together data into a single "array." Open a new script file, create your variables, and assign the data to a logically named variable. For example, if we imagine we have heights for each of four men and women, then we can do the following:

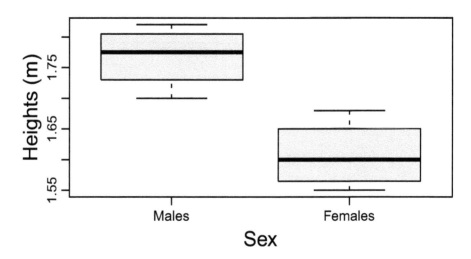

Figure 2.1 The heights of males and females, presented using the `boxplot()` function.

```
> male.hts = c(1.82, 1.79, 1.70, 1.76) # heights in meters
> female.hts = c(1.55, 1.62, 1.58, 1.68)
```

After running these lines, our data are available using these two variables by name. We can do things like get the mean or median of the heights:

```
> mean(male.hts) # calculates the arithmetic mean
```

```
[1] 1.7675
```

```
> median(female.hts) # calculates the median for continuous data
```

```
[1] 1.6
```

We also can quickly visualize these data (we'll discuss this in more detail in chapter 5). We can send these two arrays to the `boxplot()` function (see figure 2.1) like this:

```
> boxplot(male.hts,female.hts, names = c("Males","Females"),
+        xlab = "Sex",ylab = "Heights (m)", cex.lab = 1.5)
```

Note that in your script file, you do not include the line leading ">" or the "+" symbols.

2.2   Reading Data from a Spreadsheet Program

If you have data with more than about 25 values, you should enter these first into a spreadsheet and then read that file into **R**. Programs like Excel and

	A	B
1	Species	Height
2	ACSA	3.7
3	ACSA	3.9
4	ACRU	4.6
5	ACRU	5.9
6		

Figure 2.2 A sample spreadsheet with data.

Google Sheets are great at helping you organize your data and allowing you to double-check that you entered the values correctly.

You also may find data from sources such as the Centers for Disease Control and Prevention (CDC) that can be downloaded to your computer in a text format usually a "comma-separated variable" file ending with ".csv." **R** can read native Excel spreadsheets (e.g., in .xlsx format), but it is best if you save the data in the comma-separated variable format (.csv). The .csv format is highly portable and likely will be used and readable for decades. Finally, within **R** you can combine datasets and create new spreadsheets that you can save to disk and share with colleagues.

When you create a spreadsheet of data follow these rules:

1. Use descriptive, preferably one-word, unique column headings in row 1. These should start with letters but may contain numbers (see figure 2.2). Note that these will be the names of your variables.
2. Begin the data on row 2. Do not include comments or empty cells in the middle of your data, if at all possible. Variables should start with letters. Variable names can have numbers (e.g., "trmt5"). Numbers (your data) should not have characters. As with other statistics programs, if you have a column with thousands of numbers and a single word, the entire column will be considered text (you won't be able to graph the numbers or do the usual statistics on them).
3. Save the file as a text file in .csv format with a descriptive name. Note that Excel formulas will be replaced by the values. If you don't want to lose those, then save the file as an .xlsx (native) Excel file and then save a copy of it in the .csv format. The .csv format will save only the first sheet of a spreadsheet file. Avoid using multiple sheets within a single file. You can use text formats other than .csv (e.g., tab-delimited files), but reading these in is slightly different than described here. For such data, see the function read.table(), which requires you to specify

the character used to separate your data values, such as `sep = ";"` for data separated by semicolons.

4. In RStudio, change the "working directory" to where you saved the file (`Session -> Set Working Directory -> Choose Directory`). Preferably your **R** script file and data files are in the same folder so, instead, choose `Session -> Set Working Directory -> To Source File Location`. The "source file" is your **R** script file.

5. After choosing the directory, you can check that your file is in the current directory. In the console type:

```
> dir()
```

If you see your data file, you're in business. If not, set the working directory to where the file is or move your file to where you want it. You can, instead, click on the "Files" tab in the lower-right panel of RStudio. This allows you to look for files through the file structure and click on files (e.g., a `.csv` file) and view it in RStudio.

6. Read in the file using either the `read.csv()` or `read.table()` function. By default, these function return a dataframe that we assign to a variable. Below I read in a file called "`filename.csv`" that is stored in my working directory and store it in the dataframe variable called `my.data`:

```
> my.data = read.csv("filename.csv")
```

The file name must be in quotes. The variable `my.data` can be any legal variable name (names can't start with numbers and shouldn't have special characters). The variable `my.data` is now like a spreadsheet in **R** and is called a dataframe.

7. Check that the file was read in correctly and that the column headers (variables) are correct:

```
> names(my.data) # should return the names of the
> columns
```

You should see the variable names, which came from the first row of your data file. These may be different than you intended because **R** changes them a bit to be legal (e.g., spaces are converted into periods). For an overview of this important process, see box 2.1).

8. Once you are done with a variable, you can remove it from the current environment with the `rm()` function. This can be a good idea if you are going to move on to another project without closing down RStudio. To remove the variable x you simply type the following:
`> rm(x)`. You can remove all variables by clicking on the little broom

icon in the "Environment" panel of RStudio or by typing in the console the command `rm(list=ls())` and hitting `<enter>`.

If you need to change a value in your data, you should change it in the original `.csv` file using Excel. When you reread the data from disk, the new data will be stored in the dataframe, eliminating the original dataframe. Alternatively, you can make the fix in your **R** script file and be sure to run the fix after reading in the data. If you change the data in RStudio, you can write the new data to a file in your working directory using the `write.csv()` function (e.g., `>write.csv(my.data,"newfilename.csv")`).

***Box 2.1. Reading data into* R**. Here's how to read in data that are stored in a file:

1. Save the data file as a `.csv` file into the folder for your project (where your **R** script file is saved). If it's an Excel file, then, in Excel, choose "Save As...' and choose the type CSV.

2. In RStudio, set the working directory to the location of your **R** script file and the data file (click on `Session` → `Set Working Directory` → To Source File Location).

3. Read the data into a variable, for example:

```
> my.data = read.csv("file.csv")
```

4. Your variable (`my.data`) is a dataframe that holds all the data from the `.csv` file. You access the data in **R** using that variable name. The dataframe name alone points to all of the data. You can access individual columns of data using the $ sign convention (e.g., `my.data$height`). Here's what this might look like:

```
> my.data = read.csv("plant data.csv")
> my.data$heights # print heights to screen
> mean(my.data$heights) # calculate the mean of
> heights
```

We discuss dataframes more in chapter 3.

2.3   Reading Data from a Website

We sometimes rely on data that are updated frequently on a website. If we download these data, they're soon out of data, and we find ourselves having to download them again. **R** is able to read data directly from websites that provide them in a proper format. Reading the data directly from websites is best when the data are continuously being updated. Instead of download-ing the data each day, we can read data from a website with just one line of code and have the latest information. We can, for example, get the con-centration of atmospheric CO_2 at Mauna Loa, Hawaií, over the last several decades, updated weekly, at a National Oceanic and Atmospheric Adminis-tration (NOAA) website.

```
> W = "ftp://aftp.cmdl.noaa.gov/products/trends/co2/
> co2_weekly_mlo.txt"
> CO2 = read.table(W, skip = 49)[,4:5]
> plot(CO2, ylim = c(300,420), pch = 16, cex = 0.5,
+       xlab = "Year", ylab = "Atmospheric CO2 (ppm)")
```

The first line creates a text variable (W) that holds the full address for the web-site. If you go to this website, you'll see that the data file begins with a lot of information about the data. The data actually start on line 50, where we can see data columns that are separated by spaces. We use the read.table() function to get these data and store the data into a dataframe called CO2. Note that the argument skip = 49 is used to skip over all 49 lines of the header. Also, we only need the decimal date (column 4) and the CO_2 con-centration (column 5). We get this information by using the trailing [,4:5] to grab all the rows and just columns 4 and 5 (we'll learn more about how that works in chapter 3). The last line of code makes a decent graph of these data (if they're still updated at this website; you have to poke around to find the site). When we discuss visualizations in chapter 5 you will learn what the arguments in the plot() function do.

2.4   Problems

1. Below are grades earned by six randomly selected students from my bi-ology class.

$$86.3, 79.9, 92.4, 85.5, 96.2, 68.9$$

   (a) Store the grades above in a variable called grades.
   (b) Determine the arithmetic mean of the grades.

2. The following are the masses (kg) of 10 raccoons.

$$2.17, 1.53, 2.02, 1.76, 1.81, 1.55, 2.07, 1.75, 2.05, 1.96$$

   (a) Open a new file in Excel and enter "Mass" in cell A1. In cells A2 to A11, enter the data for the raccoons.

(b) Save the file as `raccoons.csv` into a working directory for your **R** projects. You need to choose a different type of file, so click on "Save as type" and choose "CSV (Comma delimited) (*.csv)". Remember where you save it! Note, to create `.csv` files in Excel, you must specify this file type under the "Save as type" option.

(c) Create a script file in RStudio named something like `coons.r` and save this file into the same folder as your `raccoons.csv` file.

(d) Set the working directory in RStudio to be the folder where your `raccoons.csv` file is stored (review the steps in box 1.4).

(e) Read this data file into an **R** variable named `my.coons`. Once you've read it in, check that the name of the variable is correct using the `names()` function. The `names()` function should return the word "Mass," which means that this variable is recognized and contains your data. If you type (`my.coons$Mass`) at the command line and hit enter, you should see your data.

(f) Summarize these data using the `summary()` function. Be sure to use the `dataframe$header` convention for your data.

3. Go to the website https://ourworldindata.org/. There you can find data on a variety of different topics. I was just checking out data on the number of COVID-19 cases in all affected countries by day (found at https://covid.ourworldindata.org/data/owid-covid-data.csv as of this writing). Ask a question about the data you have chosen and answer it by making a graph of those data. You need to copy the data you need into a new spreadsheet, save it to your computer, and then read them into a variable. Finally, make a graph of those data.

CHAPTER THREE

Working with Your Data

NOW THAT YOU HAVE DATA IN **R**, you will often need to work them a bit to get them into a form that's more usable. You'll need to know what types of data you have, know where they came from, what the units are, who collected them, and that they are correct. You also should have some sense of what the patterns in the data should look like before you analyze or graph them. This understanding can help you be more confident that you're assessing what your data might mean about the biological system you're studying.

3.1   What Kinds of Data Are There?

R is very flexible with a variety of different types of data types. You've already seen some of these. A lot of data are numeric (numbers), but they don't have to be. If they are numbers, it's often easier. Table 3.1 shows the different types of data you'll encounter and how they're represented in **R**.

3.2   Accuracy and Precision of Our Data

When working with numbers, we need to consider the difference between accuracy and precision and understand how they apply to our data (see figure 3.1). We also need to be careful about how we report our numbers.

1. **Accuracy**. When we use an instrument to measure something, we want our measurement to be as close to the actual value as possible. The closeness of this measurement to the real value is referred to as "accuracy."
2. **Precision**. This is the similarity of repeated measurements of the same value. If our measurements are quite similar, then precision is high. We can have high precision and low accuracy. Alternatively, we could actually

Table 3.1 Data Types in R.

Data Type	Description	Representation in **R**
Integers	These are whole numbers and their negative counterparts	5, −17, and 3
Real numbers	These take on any value.	5.2, 0.17, 30.3
Characters	Names of objects or treatment levels.	"cat" or "Trmt A"
Array	A one-dimensional set of the same type of objects.	`c(1, 3, 2)` or `c("a", "b", "d")`
Matrix	Generally a two-dimensional object that holds numbers.	`matrix(c(1,3,2,7), nrow = 2, ncol = 2)`
Dataframe	Like a spreadsheet of data, containing different types of data but columns have the same type of data.	`data.frame(array1, array2, array3)`
Logical	TRUE or FALSE. You will see this when you ask questions about your data.	`if (x > 0)`, then do something
Factor	This is a nominal grouping that you will use for data in dataframes	`Trmt1, Trmt2`
List	Object that contains different types of data in groups. This is a very flexible but complex type that you will rarely encounter.	`list("a",5)`
Missing data	This often happens when an experimental design is unbalanced (some individuals didn't make it in one of the treatments).	NA

have high accuracy (in one measurement) while our precision might be low (we just happened to have been lucky).

Related to these is the level of precision used to report our values, referred to as significant figures. For instance, if we we want the average mass of two feral cats, then we report that using the same number of significant digits as the mass with the least number of significant digits. If a cat is 3.0 kg and another is 5.275 kg, then the mean of those should be reported as follows:

```
> signif(mean(3.0 + 5.275),2)
```

```
[1] 8.3
```

In addition, **R** provides the `round()` function to control the number of digits reported. Below are a few examples using these two functions:

```
> a = 3.141592654 # pi
> b = 3141592654 # a x 10^9
> round(a,2) # control number of decimal places displayed
```

```
[1] 3.14
```

```
> signif(a,2) # this controls significant digits
```

```
[1] 3.1
```

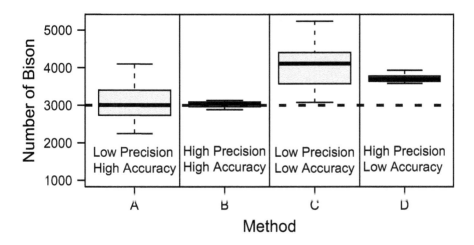

Figure 3.1 Four different hypothetical outcomes for estimates of the number of bison (*Bison bison*) in Yellowstone National Park. The true value is represented by the dashed line at 3000 animals. Each method was repeated 10 times. Method B is best, exhibiting high precision and high accuraty.

```
> signif(b,2)
```

```
[1] 3.1e+09
```

You should allow **R** to do all the calculations with all the significant digits it wants to use. However, you should report your results using the correct number of significant digits for the standards of your particular subdiscipline (or whoever is evaluating your work).

3.3   Gathering Data into Dataframes

In the previous chapter, we discussed reading data into a variable in **R** from comma-separated variable files (usually ending with .csv). When we read in a file and store it in a named variable, that variable itself points to a structure called a dataframe. Dataframes, like Excel spreadsheets, can contain many different types of data stored in different columns. Each column should have its own unique name, preferably have no empty spaces, and contain data that are all of the same type (e.g., all numbers or all character strings).

If our data are in separate variables, they can be difficult to work with. If, for instance, we had the mass of seeds from 20 separate species, then we'd have 20 different variables. Instead, a dataframe conveniently can be used to gather data together into a single variable. This is similar to a database or even an Excel spreadsheet. To do this, we send our variables to the data.frame() function and store the returned object into a single, well-named variable. Let's create a dataframe that's made up of two array variables. We'll use the male and female height data from the previous chapter:

```
> males = c(1.72, 1.59, 1.70)
> females = c(1.55, 1.62, 1.58)
```

We can combine these into a single dataframe in a variety of ways. The easiest way is to use the data.frame() function:

```
> height.dat = data.frame(males,females)
> height.dat
```

```
  males females
1  1.72    1.55
2  1.59    1.62
3  1.70    1.58
```

We now have our height data for males and females stored in a single variable called height.dat. On the left, **R** provides the row numbers. Then we see the male and female data in separate columns. If we want to see just the males we can use the $ symbol, as follows:

```
> height.dat$males # this returns an array of data for males
```

```
[1] 1.72 1.59 1.70
```

This is how we can get just one of our variables as an array. The last thing we should check out are summary statistics for these datasets. Instead of doing this separately for males and females, we can send the entire dataframe to the summary() function. Here's what happens:

```
> summary(height.dat)
```

```
     males            females
 Min.    :1.590   Min.    :1.550
 1st Qu.:1.645    1st Qu.:1.565
 Median :1.700    Median :1.580
 Mean    :1.670   Mean    :1.583
 3rd Qu.:1.710    3rd Qu.:1.600
 Max.    :1.720   Max.    :1.620
```

3.4   Stacking Data

Data often are entered into spreadsheets in many different ways, some of which are not easily analyzed. Data scientists refer to the work that goes into arranging data into a useful format as preprocessing, which can take a considerable amount of time. We will work only with data that come in two different formats. One is where data variables occur in separate columns, like we saw in the height data above for males and females. This format is called unstacked. The other format we'll use is called stacked. In general, statistics programs, like **R**, expect data to be stacked. Fortunately, it is quite easy to go back and forth if we only have one factor (in this case, "sex").

To stack the height dataframe we simply send it to the `stack()` function. This function returns a stacked dataframe.

```
> height.dat.stacked = stack(height.dat)
> height.dat.stacked
```

```
   values     ind
1   1.72    males
2   1.59    males
3   1.70    males
4   1.55  females
5   1.62  females
6   1.58  females
```

We now see that the heights are in a single column and the sexes are correctly associated with each datum in the second column. **R** assigns names to the newly created columns that aren't very informative ("values" and "ind"?). We can rename the columns like this:

```
> names(height.dat.stacked) = c("height","sex")
```

We can check that the names have changed:

```
> names(height.dat.stacked)
```

```
[1] "height" "sex"
```

We can unstack the data. You can see that **R** has used the factor levels (males and females) as the column headings:

```
> unstack(height.dat.stacked)
```

```
  males females
1  1.72    1.55
2  1.59    1.62
3  1.70    1.58
```

3.5   Subsetting a Dataframe

We often need to (and want to) work with a subset of data from a dataframe. For instance, we might need to test the normality of our height data for males and females separately. There are two basic ways to subset data. You can decide which works best for you. The two approaches are as follows:

1. Find "which" rows in a dataframe are associated with one or more search criteria (e.g., find which heights are for just females). This method returns the row indices.
2. Subset the dataframe based on one or more criteria. This returns a new dataframe with only the data you want.

When using the "which()" approach, we are seeking which rows in the dataframe have the data we need. Let's say we have the heights for males and females in the heights.dat.stacked dataframe (from above). We want just the heights for the females. The following code provides the row indices in the dataframe that contain the heights of the females:

```
> which(height.dat.stacked$sex == "females")
```

```
[1] 4 5 6
```

Note the strange double equals sign ("=="). A single equals sign is an *assignment*, while a double equals sign is read as *is equivalent to*. In English, the above command reads as "which rows in the sex column in the height.dat.stacked dataframe are females?" We can store those rows in a new variable, called "female.rows," and use that variable to get the height data for the females:

```
> female.rows = which(height.dat.stacked$sex == "females")
> height.dat.stacked$height[female.rows]
```

```
[1] 1.55 1.62 1.58
```

Alternatively, we can use the "subset()" approach to extract data from a dataframe. This function returns a new dataframe with just the information we need. Here's how we can create a new dataframe for the males:

```
> males = subset(height.dat.stacked, sex == "males")
```

We now can access the heights of males like this:

```
> males$height
```

```
[1] 1.72 1.59 1.70
```

These approaches are equally effective at getting data out of a dataframe. You should decide for yourself which approach makes the most sense to you and stick with that.

3.6   Sorting a Dataframe

There are two basic ways to sort data. The first involves the sort() function. This function returns a sorted array of numbers or characters. This can help us see the low and high values quickly, but it is not helpful if we have a dataframe with other columns of associated data. To solve that problem, we need to use the order() function. This likely will require a new way of thinking for you!

The difference between these can be better understood by looking at the height dataframe for males and females (see the stacked data in section 3.4). We can sort the heights of males and females in the dataframe using the sort() function. This function sorts an array of numbers either from lowest to highest (default) or highest to lowest (decreasing order). Let's sort the height data for males and females.

```
> height.dat.stacked$height # unsorted
```

```
[1] 1.72 1.59 1.70 1.55 1.62 1.58
```

```
> sort(height.dat.stacked$height) # sorted low to high
```

```
[1] 1.55 1.58 1.59 1.62 1.70 1.72
```

We can now see that the lowest value is 1.53 m and highest value is 1.72 m. However, we've lost the information in the dataframe that contains the sex of those individuals. To sort a dataframe with multiple columns, we need to get the correct order of the row numbers from the dataframe that will give us the sorted data. If we have the row numbers, then we can keep the sex of the individuals together with their height. To do this, we use the order() function. Admittedly, this is really confusing at first. Let's use the height data to see how this works.

```
> order(height.dat.stacked$height)
```

```
[1] 4 6 2 5 3 1
```

These are the row numbers that contain the heights in the order from lowest to highest. With this new ordering, we can rearrange the dataframe so heights increase and keep the correct sex identifiers with those height data. Here's how to sort the dataframe based on height from shortest to tallest people.

```
> height.dat.stacked[order(height.dat.stacked$height),]
```

```
  height      sex
4   1.55 females
6   1.58 females
2   1.59   males
5   1.62 females
3   1.70   males
1   1.72   males
```

This is confusing, but we can see that the data are correctly sorted from the shortest person to the tallest person. If we wanted to, we could assign that result to a new (or the same) dataframe. What's new with this syntax is the second to last character; the comma that seems wrong. Our dataframe has two dimensions: rows and columns. In **R**, like other programming languages, dataframes have index values for rows and columns in square brackets ([rows, columns]). We have provided the new order of rows but provide no value for columns (after the comma). Because we have not specified which columns to include, **R** will assume that we want all columns. This will result in the sex identifier getting ordered by row as well.

3.7   Saving a Dataframe to a File

Once we have worked our data into a form that we like, be it stacked, unstacked, or sorted, we may want to save it to a disk. Remember that in **R**, we should use script files so we write lines of code that prepare and analyze our data. We might be quite content reading in a data file that is a complete mess and fixing it up so that it's sorted and stacked, for instance, with just a few lines of code. We really don't have to do anything else with the original data because, with no additional work, we would then have our data correctly entered in **R**. Nonetheless, if you want to share these data with others, then you might want to give them a well-formatted version. To do that, you can write the dataframe to a comma-separated variable file (`.csv`) on your hard disk. Note that **R** will write the file to the current working directory. If you don't know where the file will be written, you can use the `getwd()` function at the command prompt.

Below is how you write a dataframe to a `.csv` file in the current working directory:

```
> write.csv(height.dat.stacked,"Plant heights.csv",
> row.names = F)
```

The `write.csv()` function writes the dataframe `height.dat.stacked` to the file named in quotes. Note what your working directory is (click on `Session → Set Working Directory` to control where the file will be written). The added argument `row.names = F` tells **R** to not add row numbers to the file.

3.8   Problems

1. Here are the top speeds of five cheetahs in km/hr^{-1}:

$$102, 107, 109, 101, 112.$$

(a) Enter these data into an array called `cheetahs`, in this order.

(b) Sort the data in *decreasing* order. Include all your code and the sorted data.

(c) Print out the *indices* for the data (not the data!) in the order that will result in a *decreasing* order for the data.

(d) Use those row indices to report the data in *decreasing* order. These values should be sorted in decreasing order. The answer should be the same, but your method is different. Be sure to include your code.

(e) You discover that the speed gun used to measure the cheetahs had a precision of only 10 km/hr[1] Use the signif() function to report the data properly.

2. Below are data for the masses of mussels in grams. The treatments are low, medium, and high pH levels in the water.

Low	Medium	High
12	54	87
32	34	78
22	45	59
19	69	82
27	83	64
31	44	73
25	22	77

(a) Enter the data into a spreadsheet as you see them in the table.

(b) Read the data into **R**, storing them in a dataframe variable called `mussels`.

(c) Stack the data, storing the result in the same dataframe. View them to ensure they are correct.

(d) Rename the columns "`Mass`" and "`pH.trmt.`"

(e) Sort the data in the stacked dataframe based on mass. Using the `head()` command, show the first six lines of your sorted, stacked dataframe.

(f) Get the mean (use the `mean()` function) for the `Low` pH treatment using the `which()` method.

(g) Get the mean for the `Medium` pH treatment using the `subset()` method.

(h) Create a dataframe that contains only the mussels and their treatment levels with masses greater than 80 grams. *Hint:* Use the `which()` function and a test that includes the ">" symbol.

3. Your professor wants to create working groups with the students in a laboratory. Assume there are twenty students and the professor wants five groups of four. Students count off so have ID numbers from one to twenty. Randomly place them into the five groups. Provide your code and a table with five columns and four ID numbers per group.

CHAPTER FOUR

Tell Me About My Data

AFTER WE'VE COLLECTED SOME DATA, we usually want to do something with them. We usually want to know if they tell us something important about the world. We should always inspect them to see if the values are reasonable for our system. We should ask questions about the range of the data, where the middle is, and how messy the data are. We'll first define what we mean by data because they come in many different forms. We will then look at the distribution of the data and calculate some basic statistics, or summary values, about the data.

4.1   What Are Data?

Data are usually numbers (a single number is referred to as a datum), but they don't have to be numbers (e.g., flower colors). Data can be *continuous* and take on any value in a range. We often think of height as being a continuous value because, for instance, we can really be any value between a certain range. It's not really practical, however, because we are limited by the precision of our measuring instrument. *Discrete* data have only certain values that they can be. This is true for the numbers of individuals. So count data are usually discrete.

Sometimes data come in discrete categories that can be ranked. We might not have speeds or sizes of individuals; maybe we know only the order in which the types occurred. Sometimes data have discrete categories but cannot be ranked at all. These data would simply have what we call *attributes*, referred to as *categorical* data. An example might be the color morphs of the eastern red-backed salamander. These types of data also are referred to as being *nominal* because we are simply naming different types without any ordering.

Data can be described by the type of scale they occur on. Some data are found on an *interval* scale. This means that differences between values can be compared. For instance, the difference between 32°F and 37°F is the same as the difference between 52°F and 57°F. These data, however, lack a true zero point (0°F is not a true zero). Alternatively, data on a *ratio* scale exhibit this interval property but also have a real zero point. The mass of organisms or the number of leaves on plants are on this scale. Temperature in degrees F do not exhibit this property (50°F is not twice as hot as 25°F, although I once heard a meteorologist say this). The Kelvin scale does have a zero point, however, so these values are on a ratio scale.

4.2   Where's the Middle?

There are many estimates of where the middle is in a set of numbers. The most common descriptors are the mean, median, and mode. We should recognize that these measures of where the middle is are simplifications, or models, of our data. It matters which one we choose because sometimes these measures are quite different from each other.

The Mean ($\bar{x}$)

There are actually many different means. We usually consider the mean to be what's more technically called the arithmetic mean ($\bar{x}$). This is the average you're probably familiar with, which is the sum of all the values in a sample divided by the number of observations in the sample. We can write this formally using the summation sign (Σ), which means that we add all our values from the first (i = 1) to the last (i = n) number. The numbers in the data array are represented by x_i, with i being the index for each number.

$$\bar{x} = \frac{1}{n} \sum_{i=1}^{n} x_i \tag{4.1}$$

In **R** we calculate the arithmetic mean of an array like this:

```
> x = c(5,3,6,7,4)
> mean(x)
```

```
[1] 5
```

If we have a frequency table of values, we might need the weighted mean. For instance, we might have seven ones, three twos, and five threes in our data. We might represent our data in the usually way by listing them:

$$1,1,1,1,1,1,1,2,2,2,3,3,3,3,3$$

Table 4.1 Grade table.

Grade	A	A−	B+	B	B−	C+	C	C−	D	F
Points	4	3.7	3.3	3	2.7	2.3	2	1.7	1	0
Number of grades	11	15	8	3	0	0	1	0	2	1

This is a rather short list of numbers, and we might be fine with just getting the mean using this list (equation 4.1). For long lists, however, we can get the weighted mean using the function weighted.mean(). We can calculate the weighted mean like this:

```
> vals = c(1,2,3)
> weights = c(7,3,5) # 7 ones, 3 twos, and 5 threes
> weighted.mean(vals,weights)
```

```
[1] 1.866667
```

The Median ($\tilde{x}$)

If the distribution of your data is asymmetric (skewed or lopsided; see figure 4.1) then the center of your distribution might best be described by the median, which is the 50th percentile (a percentile represents the value in your data at which that percentage of values fall below this value). If you sort your data from lowest to highest (or highest to lowest), then the median is the middle value. This is really nice if there is an odd number of values. If you have an even number of values, then the median usually is calculated as the arithmetic mean of the two values on either side of the middle. See the following examples, which use the median() function:

```
> median(c(2,3,4,5,8)) # 4 is the middle value
```

```
[1] 4
```

```
> median(c(2,3,4,8)) # the middle lies between 3 and 4
```

```
[1] 3.5
```

Let's consider an example of where we might want to find the middle of a distribution. In table 4.1 are grades, points received per grade, and the number of grades received by a graduating senior. The grades have been converted from letter grades to the often-used 0- to 4-point scale (e.g., a B− is a 2.7). Such a student's performance is often reported as a grade point average (GPA). That's the arithmetic mean.

Now, to calculate this student's GPA, we need to use the weighted.mean() function, as we did above. What is the student's GPA? This is left as Problem 7 at the end of this chapter.

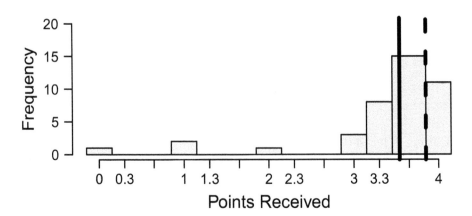

Figure 4.1 The frequency of grades received by a graduating senior. The solid vertical line represents the student's grade point average (GPA) and the vertical dashed line represents the student's grade point median (GPM).

The Mode

The mode is the most frequent observation (value) in a dataset. This makes sense if the data are discrete (e.g., integers), categorical, or nominal. If the data are continuous, however, it's likely all values in a dataset are unique. If we plot data in a histogram, we usually see some grouping (bin) that has the largest number of observations. In such a circumstance, we can think of the mode as either the range of values in this bin or as the center of the bin. This, unfortunately, is dependent on the number of bins we divide our data into. The mode in the GPA data in the previous table is an A–, and is seen as the most frequent value in the table and in figure 4.1.

In biology we rarely use the mode as a quantitative measure. Sometimes we think qualitatively about whether data exhibit two modes (e.g., bimodal distribution) or, perhaps, more modes. If the data are discrete, then the mode might be useful to use (the number of leaves on young plants). If the data are continuous, however, then this is subjective because the size and number of the bins we use affects where and how many modes we could have.

For discrete data, we can find the mode using the `table()` function. This function performs a cross-tabulation analysis, which simply gathers up all the different values and counts their occurrences. Here's an example:

```
> set.seed(100) # do this so you will have the same data
> dat.raw = sample(1:10, 500, replace = T) # get 500 random
>       # values from 1 to 10
> dat.table = table(dat.raw) # cross tabulation for the data
> dat.table # here are the data, ordered up by their frequency
```

```
dat.raw
 1  2  3  4  5  6  7  8  9 10
48 43 42 55 55 45 53 52 46 61
```

The last step is to find the maximum value (in the top row of the output above), which is our mode. To do this, we can find which value is the maximum (using the `which.max()` function). We don't want the number of occurrences but, instead, want to know which value has the largest number of occurrences. So we find which of the names in `dat.table` has the largest number of occurrences, using the `names()` function. Finally, we return that result as a number (using the `as.numeric()` function).

```
> as.numeric(names(dat.table)[which.max(dat.table)])
```

```
[1] 10
```

If our data are continuous (real numbers), then it's quite possible all our data values are unique so there wouldn't be a unique mode. However, we might look at our data using a histogram (the `hist()` function) and see that some range of values occurs most frequently in our dataset. We can actually use the histogram function to do the work for us and get which bin is the most frequent (the tallest bar in the histogram). I show the code below but without presenting the graph.

```
> set.seed(100) # do this so you will have the same data
> a = hist(rnorm(1000)) # store data from hist() in "a"
> the.mode = a$mids[which.max(a$counts)]
> cat("The mode is the bin centered on", the.mode,"\n")
```

```
The mode is the bin centered on 0.25
```

The `hist()` function is quite flexible and allows us to make as many bins as we would like (e.g., use the argument `breaks = 50`, for instance). The line above that defines the mode (`the.mode`) uses the variable "a," which is the object returned by the `hist()` function. We can use this output to find which of the bins is the largest, using the `which.max()` function. We're interested in where the midpoint is for this bin and can get that with the `mids` element from the object returned by the `hist()` function. Trust me, you don't want to do that by hand!

As you have seen, however, the number of bins in a histogram is relatively arbitrary and easily changed. Therefore, for continuous data, the mode is a subjective measure and should be avoided. Instead, we should use either the mean (symmetric data) and/or the median (symmetric or asymmetric data) to describe the middle of a distribution.

4.3   Dispersion About the Middle

After we've found the middle of our data, we next need to understand the spread of our data around this middle. Here, too, a variety of measures can be used to estimate this spread.

We're going to create a dataset that is normally distributed and then calculate a variety of measures of dispersion. If you use the code I give you, then your data will be the same as mine (using the set.seed() function). Let's first get 1,000 values from the standard normal distribution ($\bar{x} = 0$, $s = 1$).

```
> set.seed(100) # do this so you will have the same data
> x = rnorm(1000) # 1000 nums values from the "standard
>      # normal distribution"
```

Note that each of the techniques below represents a simplification, or *model*, of our data. This is similar to how we looked at different models of the center of a distribution above (e.g., the $\bar{x}$).

Range

The range of a dataset is simply the largest value minus the smallest value. It's generally a poor representation of the dispersion because it measures only the most extreme values, which we might not take much stock in. Sometimes, however, it's just what we need. We'll use this later, for instance, when we want to fit a line to some data but want to do that only over the *range* of the x-variable data (see section 9.2). We can get the range of our x variable as follows:

```
> range(x) # returns the smallest and largest values
```

```
[1] -3.320782  3.304151
```

We can use the diff() function to calculate the difference between these two values.

```
> diff(range(x)) # for this, diff() gives us what we want
```

```
[1] 6.624933
```

Standard Deviation (s)

The standard deviation is a good measure of dispersion about the mean because it has the same units as the mean. It's useful to know what the standard deviation measures. It is approximately the average, absolute difference of each value from the mean (see the section called Variance (s^2) below). If, for instance, the values are all very close to the mean, then s is small. It is easily calculated in **R** as follow:

```
> sd(x) # sd for the standard deviation
```

```
[1] 1.030588
```

Variance (s^2)

The s^2 of these data is actually the square of the standard deviation (s). The sample variance for a group of n numbers is calculated as:

$$s^2 = \frac{\sum_{i=1}^{n}(x_i - \bar{x})^2}{n - 1} \tag{4.2}$$

We can determine the variance (s^2) in **R** in either of the following two ways:

```
> sd(x)^2
```

```
[1] 1.062112
```

```
> var(x)
```

```
[1] 1.062112
```

Standard Error of the Mean (SEM)

Another measure of dispersion that is often used to represent variability is the standard error of the mean (SEM). This measure is actually an estimate of the standard deviation of a sampling distribution, which is the distribution of many means drawn from the same population. This is the foundation of the central limit theorem, which we discuss later (see the central limit theorem in figure 12.1) and is calculated as SEM = $s/\sqrt{n}$, where n is the number of values in our sample. In **R** we can calculate this as:

```
> SEM = sd(x)/sqrt(length(x))
> SEM
```

```
[1] 0.03259005
```

Some researchers choose to describe the distribution of their data with SEM. I have actually heard some people like to use this because it's smaller than the standard deviation (note that, by its definition, it will be smaller than sd because $n > 1$). This smaller value might imply that the data have a higher precision. This is not sound thinking. This measure of dispersion should be used if one wishes to show the estimated variability of the standard deviation that would be found if the means of many samples were taken. For most of us, this is kind of hard to visualize. Note that if we know the number of values in our dataset, we can get either s or s^2 from the SEM.

95% Confidence Intervals (95% CIs)

Many researchers choose to represent variability as a 95% confidence interval (see box 7.1 Fig 12.1 as before for more information). This interval is complicated, used often, and easily misunderstood. Here's how it is calculated:

$$95\% \text{ CI} = \text{SEM} \cdot t_{\alpha=0.05,df} \tag{4.3}$$

We see that this interval relies on SEM and is, therefore, an estimate of a population parameter that assumes we sample our population many times. In the equation, SEM is followed by a value drawn from the t-distribution, which (statisticians, look away) can be thought of as a measure related to the normal distribution. The 95% comes from $1 - \alpha$. The *df* represents the degrees of freedom, which is the number of observations minus one. Here's how to calculate the 95% confidence level for our data using **R**:

```
> n = length(x) # the size of a sample
> CI95 = SEM * qt(0.975, df = n - 1)
> CI95
```

```
[1] 0.06395281
```

The qt() function returns a t-value, given that we want a two-sided distribution with n observations (that's where the 0.975 comes from: $0.975 = 1 - \alpha/2$). We'll talk more about this in chapter 7.

Using SEM in this way suggests that we are somehow estimating some measure of dispersion from a large number of samples. The 95% confidence interval is a measure that, if we calculated this interval for many samples, 95% of those intervals would capture the true population mean. It is wrong to think of this as an interval that says we're 95% confident that the true mean lies in this range. The key to the correct definition lies in its reliance on SEM. All this depends on the assumption that our data are perfectly normally distributed and on the central limit theorem (see chapter 12 in figure 12.1).

You probably noticed that the measures s, s^2, SEM, and the 95% CI rely on $\bar{x}$ (e.g., see equation 4.2 for the variance, which relies on $\bar{x}$). This suggests that, for these measures of dispersion to be useful, the $\bar{x}$ must be a good measure of the middle of our data. These measures also weigh all deviations from the $\bar{x}$ (either smaller or larger) equally. Therefore, the utility of these dispersion measures relies on the assumption that our data are *normally distributed*. We will learn about these parametric statistical tests in later chapters.

The Interquartile Range (IQR)

The last measure of dispersion I want to mention is the interquartile range (IQR). This is a range that is based on quartiles, like the median, which is the second quartile or the 50th percentile. The IQR is the third quartile (the 75th percentile) minus the first quartile (the 25th percentile). The IQR is a measure of dispersion around the median. Here's how we calculate this in **R**:

```
> IQR(x) # returns the interquartile range of the x array
[1] 1.359288
```

This calculation can be confusing for small datasets, so it should be considered an approximation. If the data are continuous and have a large number of values, then the measure makes sense.

The IQR is not often used in biology, but occasionally we are interested in the boundaries on the middle 50% of observations when the distribution is not necessarily normal. My college uses this range of the SAT scores for entering students by reporting the 25th and 75th percentiles.

Be sure to check with whomever is evaluating your work to find out which measure of dispersion is preferred.

The coefficient of variation (CV)

The Coefficient of Variation (CV) is a valuable measurement of variability when we're comparing objects that are different. Let's imagine we have measured the mass of 10 mice and 10 elephants. We might ask, "Which of these organisms exhibits more *relative* variability?" Clearly the elephants have greater *absolute* variability in kilograms, but maybe the mice are actually more different from each other than are the elephants from each other. The CV is usually expressed as a percentage and is calculated as follows:

$$CV = 100 \cdot \left(\frac{s}{\bar{x}}\right) \tag{4.4}$$

where s is the standard deviation and $\bar{x}$ is the mean.

4.4 Testing for Normality

We've already talked about data being normally distributed. We're now going to discuss this more explicitly and learn how to test whether our data are normally distributed. As we discuss this, we should keep in mind that no biological data are truly normal in the statistical sense because the mathematical distribution for a normal curve ranges from minus infinity to plus infinity. What that means for heights or masses is, obviously, questionable in real biological systems.

So we will test whether the data are significantly different from a normal distribution before performing a statistical test. As mentioned above, the reason for this is that some statistical tests, called parametric tests, assume that the arithmetic mean of the sample is a good measure of the middle of the data and that the spread of the data is well represented by the standard deviation. If the mean is not in the middle of the distribution and/or the standard deviation poorly represents the spread, then the assumptions of parametric tests are violated (e.g., the statistical test and, therefore, your conclusions could be terribly wrong). Non-parametric statistical tests do not make the assumption that the distribution of the data is normal. These non-parametric tests do assume, however, that multiple samples, though not normally distributed, have the same distributions (because our null hypothesis is that they come from the same population).

The bottom line is that you must test whether your data are normally distributed before doing a statistical test. Some people think if you're not sure whether the distributions of your samples are normally distributed, you should just use non-parametric tests. This is not a safe approach and may lead you to make mistakes in your interpretation of your results. In short, always use the correct statistical test!

As we will see, parametric statistical tests include many of the tests with which you are familiar, including regression, correlation, t-test, and analysis of variance tests. Non-parametric tests include the Mann-Whitney U and the chi-square tests.

We may, at times, be able to *transform* our non-normal data into normal data so that we can employ parametric tests. We might do this with growth data, for instance, because many organisms and populations grow exponentially. Log-transforming such data can lead to normally distributed data (see section 4.6).

Our first step to assess a dataset's distribution is to create a graph of the data. Three common graphs for this purpose are histograms, boxplots, and Q-Q plots (see figure 4.2). All three of these plots are conveniently graphed for us using the function `simple.eda()`, found in the package `UsingR`. The "eda" stands for exploratory data analysis. Be sure to install that package, if you haven't already, and load it using the `library()` function (see section 1.6).

```
> simple.eda(rnorm(1000))
```

All three visualization approaches can help us evaluate whether our data are normally distributed. However, I encourage you also to use the Shapiro-Wilk test (the `shapiro.test()` function) to help you decide whether data are normally distributed. The null hypothesis for normality tests is that the

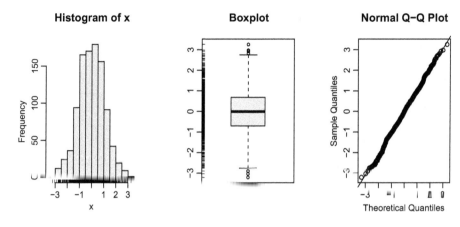

Figure 4.2 Exploratory data analysis visualizations for 1,000 values drawn from the standard normal distribution using the function `simple.eda()` from the `UsingR` package. The left graph is a histogram. The center graph is a modified boxplot. The right graph is a Q-Q plot. If the data points in the right graph fall on the straight line, then the data adhere to a normal distribution.

data *are* normally distributed. This means that if we get a p-value greater than 0.05, we do not have enough evidence to reject the assumption of normality (see section 6.8 for a complete explanation of p-values).

Let's generate four samples ($x1 - x4$) and evaluate their distributions for normality. Here are the four datasets (if you run these lines of code, you should have the same data):

```
> set.seed(10) # do this so you will have the same data
> x1 = rnorm(1000) # 1000 values from standard normal dist.
> x2 = runif(1000) # 1000 values from uniform dist.
> x3 = rgamma(1000,2,1) # 1000 values from a gamma distribution
> x4 = c(rep(3,10),1,2,4,5) # leptokurtic distribution
```

Before we test these datasets for normality, it's always a good idea first to look at their distributions. I've done this by sending each dataset to the `hist()` function (see figure 4.3).

The upper-left panel of figure 4.3 looks like a bell-shaped curve, which is indicative of a normal distribution. The other three, however, seem quite different from a normal distribution. Let's evaluate these samples for normality using the Shapiro-Wilk test for normality.

```
> shapiro.test(x1)$p.value # large p -> fail to reject Ho
```

```
[1] 0.25377
```

```
> shapiro.test(x2)$p.value # small p -> reject Ho
```

```
[1] 2.90682e-19
```

Normal Distribution

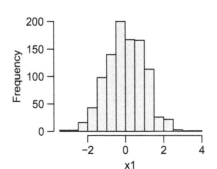

Uniform Distribution

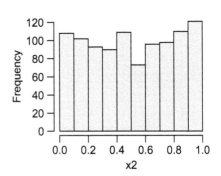

Gamma Distribution

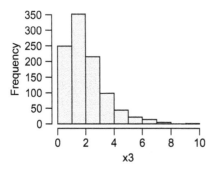

Leptokurtic Distribution

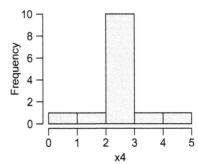

Figure 4.3 Histograms for four distributions for the datasets x1, x2, x3, and x4. We fail to reject the normality null hypothesis only for the distribution in the upper-left panel (the other three are not statistically normally distributed). Note that the data in both lower panels exhibit kurtosis, which is referred to as being leptokurtic, meaning they are peaked. The normally distributed data are mesokurtic, while the data in the uniform distribution are platykurtic, or flat-topped.

```
> shapiro.test(x3)$p.value # small p -> reject Ho
```

```
[1] 2.789856e-25
```

```
> shapiro.test(x4)$p.value # small p -> reject Ho
```

```
[1] 0.001854099
```

The output from the shapiro.test() function suggests that only the x1 dataset is normally distributed ($p = 0.254$) while the results for x2, x3, and x4 suggest that these datasets are not normally distributed ($p \leq 0.05$). This agrees with our original interpretation of the histograms in figure 4.3.

If the data are *not* normally distributed it can be useful to know in what way they violate normality. You've probably heard of a distribution appearing to be skewed. This is when the data exhibit an asymmetric distribution (see the lower-left panel of figure 4.3). We can test skewness using the `skewness()` function from the `e1071` package. Be sure you've installed the `e1071` package (see section 1.6) and loaded it with the `library()` function. Now you can run the following lines of code to test for skewness.

```
> skewness(x1)
```

```
[1] -0.008931715
```

```
> skewness(x2)
```

```
[1] -0.03822913
```

```
> skewness(x3)
```

```
[1] 1.373072
```

```
> skewness(x4)
```

```
[1] 0
```

Skewness values close to zero suggest no skew in the data. We see this for the datasets x1, x2, and x4 in figure 4.3. However, the x3 (lower-left panel of figure 4.3) is positively skewed. We say these data are skewed to the right (the direction the tail is pointing). The uniformly distributed data (upper-right panel of figure 4.3) are not skewed, so they are probably not normally distributed due to our next measure of kurtosis.

Kurtosis, our last measure of the deviation from normality, is a measure of the distribution's shape away from the standard bell-shaped distribution. You can try this for all four distributions as follow:

```
> kurtosis(x1)
```

```
[1] -0.1145157
```

```
> kurtosis(x2)
```

```
[1] -1.296026
```

```
> kurtosis(x3)
```

```
[1] 2.356125
```

```
> kurtosis(x4)
```

```
[1] 1.104286
```

Table 4.2 Statistical moments about the mean.

Statistic	x_1	x_2	x_3	x_4
$\bar{x}$	0.011	0.508	1.975	3
s^2	0.984	0.09	1.946	0.769
Skew	−0.009	−0.038	1.373	0
Kurtosis	−0.115	−1.296	2.356	1.104

If the value of kurtosis is near zero, then the distribution is mesokurtic, or it is consistent with a bell-shaped curve without a really sharp peak or a flat-topped peak. If the value is a relatively large, positive number, then the data are more pointy in the middle than a regular, bell-shaped distribution ($x3$ and $x4$). Such a distribution is referred to as being leptokurtic. If the value is negative, then the distribution is relatively flat-topped. As we can see here, the kurtosis value for $x2$ is more negative than the $x1$ sample. The $x2$ distribution is referred to as being platykurtic.

To summarize, we can look at these four statistics, called the first four moments about the mean, and compare them for the four distributions in table 4.2:

4.5 Outliers

We all know that outliers are unusual values in a dataset. It turns out they're kind of hard to define, and we'll do so formally later in chapter 5. Outliers are rare (they should occur in less than 1% of values drawn from a normal distribution), but they are expected to occur if our sample size is large enough. I've heard some people say that outliers should simply be thrown out. This is wrong. You should never discard data simply because the values seem unusual; we need a better reason than that! If a value seems to be totally unexpected then you should investigate why.

Here are some possible explanations for why a data point might appear to be an outlier or otherwise questionable:

1. Equipment malfunction.
2. Recording error.
3. Data entry mistake.
4. General carelessness.
5. Extreme environmental conditions at the time of the data collection.
6. A large sample size, which will yield statistical outliers (they're expected).

If, after careful consideration, you decide that a value is an outlier that either needs to be removed from your dataset or needs to be changed (e.g., it should be 1.53 not 153), then here's what you should do. If the data were read into **R** from a file, then you should fix the data file and read it back in, if possible. If this is not possible, then you need to find it using **R** and fix it. Let's imagine you have a dataset of 100 values, they're all $1.0 \le x \le 2.0$, and one of the values is 153 (it's supposed to be 1.53). You can find its index value in the array (where in the list of 100 numbers it is found). First, we need a list of 100 numbers in our range:

```
> x = runif(100)+1 # creates 100 random values (1 ≤ x ≤ 2?)
```

Next, let's change the thirty-sixth value to 153:

```
> x[36] = 153 # makes the thirty-sixth value erroneously 153
```

This assigns 153 to the thirty-sixth array element. Finally, let's find that value and fix it. Here's how we can find the index of any value in an array greater than 2:

```
> which(x > 2) # find index of number(s) that are > 2
```

```
[1] 36
```

R tells us that, in this case, the outlier is the thirty-sixth entry in the x array. We can inspect that value as follows:

```
> x[36]
```

```
[1] 153
```

and see that the thirty-sixth element is out of range. We double-check our lab notebook and see that it's supposed to be 1.53. Here's how you can fix it:

```
> x[36] = 1.53
```

What if the number was recorded correctly, but it seems our instrument gave us a faulty value? Then we might want to simply remove this value. If you want to remove the value from the working copy of data in **R**, you can do the following, leaving you with just 99 data points:

```
> x = x[-36] # -36 means removes the 36th value, then copies
>            # the remaining values back into x
```

Remember, if the data are in a spreadsheet, you should fix the spreadsheet and read the data back into **R**.

4.6   Dealing with Non-Normally Distributed Data

If the data are not normally distributed, what can you do? Before jumping to a non-parametric test (a statistical test that does not assume data are normally distributed but are, generally, weaker tests of hypotheses), you should investigate whether you can transform the data so they become normally distributed. This can be a good idea because, if the data are normally distributed upon transformation, a parametric statistical test can and should be used. The exponentially distributed data in the histogram in the left panel of Figure 4.4 are shown without transformation and are then log-transformed, using the natural logarithm. Many growth processes in biology lead to exponential distributions that are normalized with this log-transformation.

Below is the Shapiro-Wilk test for normality for these two datasets. Recall that the null hypothesis for this test is that the data are normally distributed. Therefore, if $p > 0.05$ then the data appear to be normally distributed (i.e., we don't have enough information to reject the H_0).

```
> shapiro.test(Data) # these are NOT normally distributed
```

```
        Shapiro-Wilk normality test

data:   Data
W = 0.90762, p-value = 3.286e-06
```

```
> shapiro.test(log.Data) # transformed -> normally distributed
```

```
        Shapiro-Wilk normality test

data:   log.Data
W = 0.98908, p-value = 0.5911
```

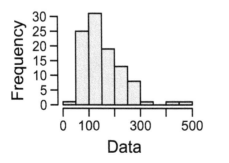 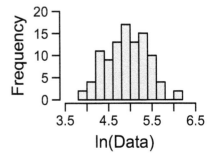

Figure 4.4 Two histograms of 100 data points. The left histogram shows data that are right-skewed. The histogram on the right shows the same data after log-transforming. The data are now normally distributed (a log-normal distribution). The transformation was done using the natural logarithm function (`log()`).

Sometimes you have data with perhaps two or more groups and discover that one sample is normally distributed while the other is not. You can try to transform the data and the samples may be normally distributed. What might happen, however, is that the non-normal data may become normally distributed and the previously normal data may become non-normally distributed. What should you do? This is tricky so discuss this with a statistician or adviser. What can become critically important in such cases is knowing what the underlying distribution should be or knowing how previously collected data were distributed. If the data are usually normally distributed, or they should be for a very good reason, then you might assume normality.

4.7 Problems

1. The following values represent the average lengths of flagella in *Chlamydomonas* (a green alga) as a function of the number of flagella per cell (based on data from Marshall et al., 2005). If the alga has one flagellum, then, on average, they are 11.1 μm in length. They are only 4.3 μm, on average, when they have six flagella. Store these in a variable called flag.

11.1, 11.0, 10.7, 10.9, 11.2, 11.2

   (a) Calculate the mean, standard deviation, median, SEM, IQR, range, and CV for these data.
   (b) Which of the measures of central tendency and variability assume that the data are normally distributed and which do not?
   (c) Create a histogram, boxplot, and Q-Q plot with a line using the function simple.eda() from the UsingR package.
   (d) Test the data for normality using the Shapiro-Wilk test. Does this statistical test seem to support what the visualizations suggest? What concerns do you have about this analysis?

2. Create a set of 1,000 normally distributed random numbers (use the rnorm() function) with a mean equal to your height in cm. Assume $sd = 12$cm. Store these in a variable called heights.

   (a) What is the mean height for your dataset? *Hint:* It should be close to your own height.
   (b) Report the values of any outliers. *Hint:* Use the output from the boxplot() function.
   (c) Create a histogram of your data.

3. The human population has been growing rapidly over the past 2,000 years. From the internet, find the current size of populations for the top ten most populous nations.

   (a) Create a histogram, boxplot, and Q-Q normality plot of these data using the `simple.eda()` function.
   (b) Describe the distribution of population sizes across these nations.

4. Provide one, clear biological example for each of the following data types not discussed in this chapter:

   (a) Continuous.
   (b) Discrete.
   (c) Categorical.

5. Can you think of an example other those discussed in this chapter where you'd need the range of the data?
6. I collected the mass of five mice and five elephants in the table below. All data are in kilograms. Which exhibits the greatest relative variability? *Hint:* Use the CV from section 4.3.

Mice	24.0×10^{-3}	22.0×10^{-3}	10.0×10^{-3}	18.0×10^{-3}	27.0×10^{-3}
Elephants	5.19×10^{6}	4.40×10^{6}	4.82×10^{6}	4.19×10^{6}	4.87×10^{6}

7. What is the GPA of the student whose grades are provided in table 4.1? Be sure to provide the data you enter and your code, using the `weighted.mean()` function.

CHAPTER FIVE

Visualizing Your Data

IN THIS CHAPTER, we will work to create clear, concise graphs of relationships to convey information efficiently to your readers. **R** is extremely powerful and flexible in allowing us to create a variety of visualizations, from simple scatterplots to full-blown animations and three-dimensional (3D) clickable diagrams. With its basic drawing palette, you're actually able to create any visualization you are likely to need or see in any professional scientific publication. In this chapter we will develop a variety of graphs, enhance those graphs, and discuss the reasons why we might use each of these graph types.

As we begin, you will undoubtedly think **R** is rather primitive. Writing a line of code to generate a graph might even seem crazy. But you will likely soon realize that it is easy and intuitive to use typed words that direct **R** to use your data to make the visualizations you want. Yes, it might seem hard at first, but I hope you enjoy the skills you develop to make publication-quality graphs. To see even more of what's possible using **R**, you can search for examples online and, perhaps, begin by checking out this site: https://www.r-graph-gallery.com/.

In this chapter, you'll see a variety of basic graphing procedures. The more you play with the different graphs, the better you'll get. Be sure to give them a try.

5.1 Overview

Before we get started, let's create some random data. The following lines of code will create three variables (x, y, and z) that each have 100 values drawn from normal distributions with standard deviations equal to 1. The first line sets the random number seed so that your data will be the same as my data. This way your graphs should look exactly like those I've created.

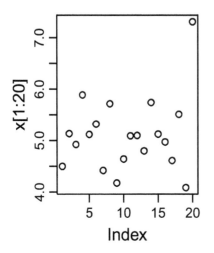
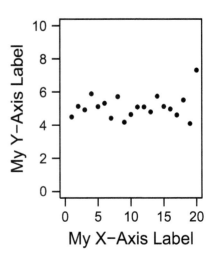

Figure 5.1 An example of a default scatterplot (left) and one with modifications (right). For the graph on the right, I've added arguments that make it more professional looking (see box 5.2).

```
> set.seed(100) # do this so you will have the same data
> x = rnorm(100, mean = 5) # 100 random nums, mean = 5, sd = 1
> y = rnorm(100, mean = 6) # 100 random nums, mean = 6, sd = 1
> z = rnorm(100, mean = 8) # 100 random nums, mean = 8, sd = 1
```

As you proceed through this section you'll see a number of different types of graphs. Graphs in **R** are made by sending data variables to graphing functions. We refer to these data, and anything else we send to a function, as arguments. In the examples that follow, you will see how data are sent to the graphing functions and a variety of optional arguments that improve the looks of the graphs. Keep your eyes open for when I do this in the examples. Some of the important arguments used to improve the looks of graphs are summarized in box 5.2 at the end of this chapter. You can see how this can help improve the scatterplot in figure 5.1.

```
> par(mfrow = c(1,2)) # make graph window 1 row and 2 columns
> plot(x[1:20]) # left plot with no graphics parameters
> plot(x[1:20], xlim = c(0,20),ylim = c(0,10), las = 1,
+      xlab = "My X-Axis Label", ylab = "My Y-Axis Label",
+      main = "My Custom Title", pch = 16, cex = 0.75,
+      cex.lab = 1.5) # a professional-looking graph!
```

Legends

In addition to the graphical parameters just listed, you might be interested in adding a legend to your graph. You'll see that some graphing types, like `barplot()`, can accept a `legend.text` argument (see figure 5.6). For other graphs, you can always add a legend using the `legend()` function. Because this is a function, it is called separately, on its own line. This can be a little tricky and so you might just steal some code from later in this chapter or from online sources and tweak the code for your purpose. There's an example of the implementation of this within a scatterplot in figure 5.7.

The `legend()` function is a little tricky. The first argument is where the legend should be placed. Commonly we'll place it in the top left (`"topleft"`) or top right (`"topright"`) of the graph. It often takes up too much room, so you need to increase the scale for the y-axis. One last tweak I have found helpful is to make the legend horizontal by adding the argument (`horiz = T`). The words you want to identify the different samples are combined into an array and passed to the `legend` argument (e.g., `legend = c("Females","Males")`). You can then add color and fill for lines and boxes as needed.

5.2 Histograms

Histograms are standard graphs for visualizing the *distribution* of a dataset. This is the best plot for getting a first look at your data. It's quite easy, for instance, to see if the data appear normally distributed, which is important to know as you begin testing hypotheses. Below is the code used to make the side-by-side histograms in figure 5.2.

```
> par(mfrow = c(1,2))
> hist(x, cex.lab = 1.5, main = "", las = 1)
> abline(v = mean(x),lwd = 5)
> hist(x, cex.lab = 1.5, main = "", breaks = 34,
+       las = 1, ylim = c(0,10))
> abline(v = mean(x),lwd = 5)
> par(mfrow = c(1,1))
```

You can control the number of bins that your data are placed into with the "breaks" argument (e.g., `breaks = 34`). You might first simply let **R** make the histogram and see how it looks before changing the number of bins. If the data are normally distributed, they should form a bell-shaped distribution.

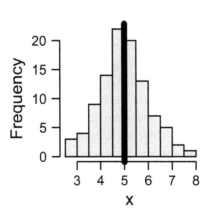

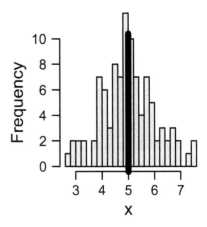

Figure 5.2 Two histograms of the *x* data. On the left, I let **R** select the number of bars to graph. The thick, vertical line is placed on the graphs at the mean using the `abline()` function. For the histogram on the right, I requested thirty-four bars (`breaks = 34`).

5.3   Boxplots

Boxplots, like histograms, can be used to show the distributions of data. They are preferred over barplots because they show the distributions of data and make comparing distributions among samples easy and clear.

Boxplots generally show a box with a line through the middle. The top of the box is the 75th percentile, the middle line shows the median, which also is the 50th percentile, and the bottom of the box is the 25th percentile. The range between the 75th and 25th percentiles is referred to as the interquartile range (IQR). A percentile is a value at which that percentage of observations is below. Therefore, the 50th percentile is the value at which 50% of the observations are *below*. A perfect score on an SAT test usually represents *only* the 95th percentile because 5% of those who take the exam get perfect scores. The whiskers may extend above the 75th percentile to the largest value but must not exceed 1.5 times the interquartile range above this 75th percentile. There also can be a whisker below the 25th percentile in the same manner. Data points that lie beyond the whiskers are called outliers (see box 5.1).

Box 5.1. *What is an outlier?*
An outlier is a value that is more extreme than 1.5 times the interquartile range above or below the 75th or 25th percentiles, respectively. Or, more simply, they are the points beyond the whiskers of a boxplot when graphed using **R**.

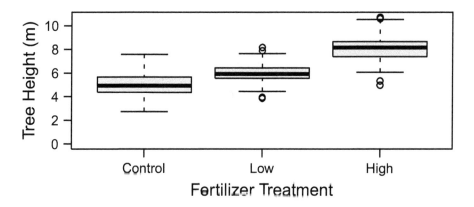

Figure 5.3 A boxplot of tree growth under three different conditions (A, B, and C). Note that the y-axis extends down to zero (`ylim = c(0, ...)`).

In figure 5.3 I've plotted *x*, *y*, and *z*, pretending they are the heights for three groups of trees grown under different conditions (treatments). Treatment levels B and C both have outliers.

```
> boxplot(x,y,z,names = c("Control","Low","High"),
+         xlab = "Fertilizer Treatment", las = 1,
+         ylab = "Tree Height (m)",
+         cex.lab = 1.5, ylim = c(0, max(c(x,y,z))))
```

5.4   Barplots

Barplots generally show a summary value for continuous variables. The heights of the bars almost always represent the arithmetic means. Sometimes bars are displayed horizontally, but this is usually not best because the response variable (the variable that was measured) should be shown on the y-axis. The function `barplot()` needs just the heights of each bar in a single array. In the example, I use the means, placed into a new variable called Ht (see figure 5.4). Note that barplots should extend down to zero on the response variable axis (usually the y-axis).

```
> Ht = c(mean(x),mean(y),mean(z))
> barplot(Ht,
+         xlab = "Fertilizer Treatment",
+         ylab = "Tree Height (m)", las = 1,
+ names = c("Low","Medium","High"), cex.lab = 1.5)
> abline(h=0)
```

Be careful when using a barplot. You've seen these often, but they represent a very simplified version of your data (the mean is a single datum) as

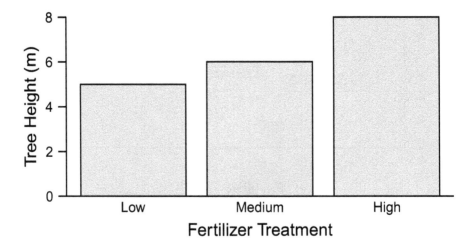

Figure 5.4 A barplot showing the means of individuals of a tree species grown under three different levels of fertilizer. These are the same data as those shown in figure 5.3. Does something seem to be missing?

opposed to boxplots, which show several pieces of information. If you use barplots, you should consider including error bars that represent the variability of the data (discussed in box 7.1). Error bars on barplots still are limited, reducing your data to only two pieces of information (the mean and the estimate of variability). In addition, if you graph the means using a barplot be sure that they are good estimates of the center of distributions (i.e., the data should be normally distributed).

Let's compare the differences between a plain barplot and a boxplot for two samples of data (A and B) with the same means and variabilities (figure 5.5). The datasets are the same for both graphs. You can see that the barplot on the left, showing only the mean ± the standard deviation, suggests the samples are very similar, while the boxplot on the right shows that the two datasets are wildly different. This is a very important reason for using boxplots instead of barplots, when possible. The reason the barplot is the wrong graph type to use is because the data in treatment B are far from normally distributed. This is easily seen in the boxplot in the right panel of figure 5.5.

Sometimes we want to create a barplot that compares observations over two factors. This is often used when we have data we are analyzing using an analysis of variance (see chapter 8) or a two-way chi-square test (see chapter 10). If we want a barplot, then we need to decide how to gather up our data. Let's imagine we have the following average Medical College Admission Test (MCAT) scores for females and males who either are or are not majoring in biology.

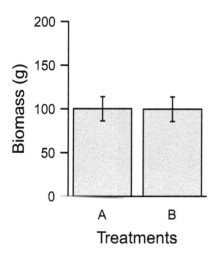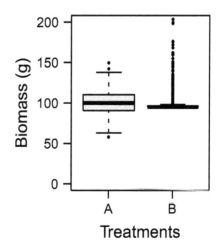

Figure 5.5 On the left is a barplot showing means (± standard deviations) for two samples of plants (treatments A and B). These treatments appear to have resulted in no differences in plant mass. The boxplots on the right show the same data but reveal that the samples are really quite different from each other when we take into account the distributions of the data, despite having nearly identical means and estimates of variabilities. Data for samples A and B are the same in both plots.

If we wish to plot these means using a barplot, we have to decide which factor goes on the x-axis and which factor is represented by the trace factor (that's the factor that's paired and shown in the legend). The factors can be interchanged, and you'll have to decide which one makes the most sense for your data. For these data, I have done it both ways (see figure 5.6). Here are the data and how they are combined into a matrix for the `barplot()` function.

```
> females = c(36.3,29.5)
> males = c(36.2,28.9) # first is biology, then non-bio majors
> MCAT = matrix(c(females,males),byrow = T, nrow = 2)
```

To create the side-by-side barplots, we need the data to be in a matrix. The command above stores the data as a matrix in the variable MCAT. The default for the `matrix()` function is to combine the data by columns. I like to enter the data by rows, and so I added the argument `byrow = T`. Let's see what the data look like:

```
> MCAT
```

```
      [,1]  [,2]
[1,]  36.3  29.5
[2,]  36.2  28.9
```

This matrix matches the data in table 5.1. The code below creates figure 5.6.

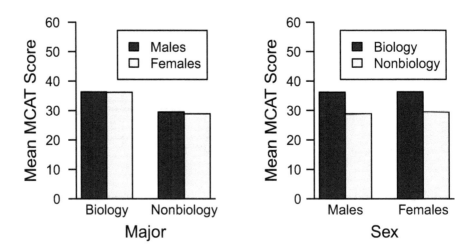

Figure 5.6 Two barplots that represent the same data in two different ways. The graph on the left shows mean MCAT scores for biology versus nonbiology majors, separated by sex. The graph on the right shows the same data for males and females, each separated by major. (Note that the scores were conceived by me for my biology audience.)

```
> par(mfrow = c(1,2)) # graphics panel with 1 row, 2 columns
> MCAT = matrix(c(females,males),byrow = T, nrow = 2)
> barplot(MCAT,beside = T, names = c("Biology","Non-Biology"),
+        ylim = c(0,60), xlab = "Major", cex.lab = 1.5,
+        las = 1,
+        ylab = "Mean MCAT Score", legend.text = c("Males",
+        "Females"))
> abline(h=0)
> # Alternatively, separate by major
> bio = c(36.2,36.3) # first males, then females
> non.bio = c(28.9,29.5)
> MCAT = matrix(c(bio,non.bio),byrow = T, nrow = 2)
> barplot(MCAT, beside = T, names = c("Males","Females"),
+        ylim = c(0,60), xlab = "Sex", las = 1,
+        cex.lab = 1.5, ylab = "Mean MCAT Score",
+        legend.text = c("Biology","Non-Biology"))
> abline(h=0)
```

Table 5.1 Average MCAT scores for males and females by major.

Sex	Biology	Nonbiology
Females	36.3	29.5
Males	36.2	28.9

5.5   Scatterplots

Scatterplots are used to show the relationship between two, usually continuous variables. You might just show points for viewing relationships, or you might add a best-fit line to a scatterplot (see section 9.2 on linear regression) when there's a statistically significant dependency that you want to model. Such plots are made with the plot () function, which takes at least data for the y-axis variable. In addition, we usually provide the function x-axis data as well. We saw these types of graphs in figure 5.1.

Scatterplots with Different Points and a Legend

We may have several different sets of data we want to show on the same graph. We can easily add points and lines using the function points () and lines (), respectively.

Below are some data I retrieved from a paper by White and Seymour (2003) on the relationship between basal metabolic rate and mass of different animals. I use only a subset of data from the deer Order and the carnivore Family. I have formatted the graph using log-log axes (argument log = "xy"), added exponentiated labels for the scale, and provided exponential and subscript values in the y-axis label. This really is not easy, so you might re-create this graph and keep the code for later use (see figure 5.7).

```
> par(mar = c(5.1,5.1,2.1,1.1))
> M.artiodactyl = c(37800, 196500,69100, 325000,21500,58600,
> 20500)
> BMR.artiodactyl = c(9318,41242,19120,51419,8308,25609,5945)
```

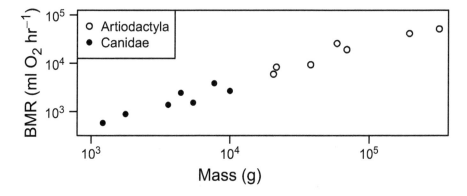

Figure 5.7   The relationship between basal metabolic rate (BMR) and the mass of animals from the deer and carnivore groups. Note that the many elements to this graph make it complicated. You should consider copying the code that generated this graph and save it to a file. Data from White and Seymour (2003).

```
> plot(M.artiodactyl, BMR.artiodactyl,
+      xlim = c(1000,300000), ylim = c(400,100000),
+       log = "xy", cex.lab = 1.5, xaxt = "n", yaxt = "n",
+        xlab = "Mass (g)",
+       ylab = expression(paste("BMR (ml O"[2]," hr"^-1,")")))
> axis(2,at = c(1e3,1e4,1e5),
+        labels = expression("10"^3,"10"^4,"10"^5),las = 1)
> axis(1, at = c(1e3,1e4,1e5),
+        labels = expression("10"^3,"10"^4,"10"^5))
> M.canids = c(3600,10000,7720,5444,1215,1769,4440)
> BMR.canids = c(1374,2687, 3860,1524,583,887,2442)
> points(M.canids, BMR.canids,#log = "xy",
+          pch = 16, las = 1)
> legend("topleft", legend = c("Artiodactyla","Canidae"),
> pch = c(1,16))
```

5.6   Bump Charts (Before and After Line Plots)

Sometimes we have data with measurements on subjects before and after applying a treatment. For instance, we can measure reaction time to a stimulus in a variety of organisms. We then train or expose subjects to a treatment and again measure their responses. We can make a nice graph in **R** by following these steps:

1. Enter the data into two array variables of equal length (e.g., before and after). Alternatively, enter the data into a spreadsheet with two columns of equal length. Each row would represent the response of a single subject.
2. Create an empty plot.
3. Draw lines that connect the paired data (before and after).
4. Draw points at the ends of each of the lines, if we want.

Below are lines of code that create the data and then graph them as a bump chart (see figure 5.8).

```
> N = 20 # number of individuals in sample
> time1 = rnorm(N, mean = 3)
> time2 = rnorm(N, mean = 7)
> plot(0, xlim = c(0.75,2.25),ylim = c(0,10),
+       type = "n",xaxt = "n", xlab = "Time",
+       ylab = "Measurement", cex.lab = 1.5, las = 1)
> axis(1,at = c(1,2),labels = c("Before","After"),
+       cex.axis = 1.25)
> for (i in 1:length(time1)) { # add the lines one at a time
+   lines(c(1,2),c(time1[i],time2[i]),lty = 1)
+ }
> points(rep(1,length(time1)),time1, pch = 16, cex = 1.5)
> points(rep(2,length(time2)),time2, pch = 17, cex = 1.5)
```

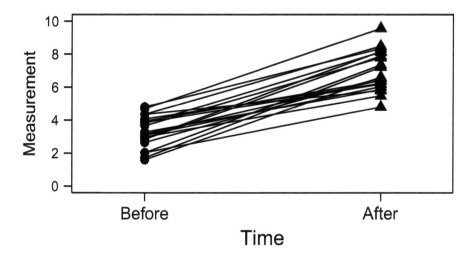

Figure 5.8 A bump chart showing lines that connect observations made on individuals before and after some treatment. We provide a graph like this so that a reader might see an overall trend between subjects. Notice that each *line* represents a measurement from a *single* experimental unit, so these data can't be analyzed as though the before and after points represent independent samples (see section 7.4).

The empty plot is created by including the argument "`type = "n"`" and then use a "`for`" loop to fill in each line one at a time. This looping is explained in more detail in chapter 12. The resulting graph is figure 5.8. A good time to use this type of graph is seen later in section 7.4.

5.7   Pie Charts

A pie chart is a type of graph that might be used when data values add up to 100%. However, simply stated, pie charts are rarely used in biology. They are challenging because the amounts in each slice often are too hard to compare quantitatively. To overcome this problem, pie charts often display the numerical value of the size of the slice, as well. However, if a graph requires the presence of data values, then the pie chart is probably not appropriate. If there are few slices in the pie then those data might best be presented in a table.

For instance, we can look at the data presented at the top of figure 5.9. The two charts show different data, but it is difficult to distinguish the differences. As you look at these pie charts, your eyes will go back and forth trying to see if there are any important differences. Alternatively, the data are better represented using another graphing procedure, such as a barplot (lower-panel, figure 5.9), which more clearly shows differences between groups.

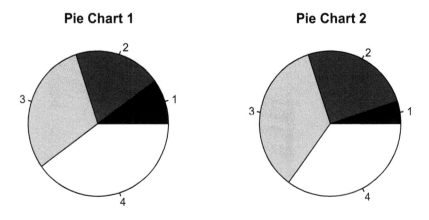

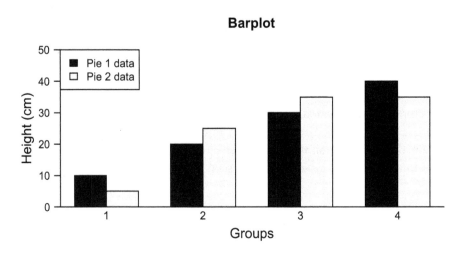

Figure 5.9 The upper panel shows two pie charts with slightly different data. The differences between each group across the two pie charts are difficult to see. You never want to make your reader work to understand your results. These data are presented more clearly in a barplot such as the one shown below the two pie charts. The two datasets are graphed using a side-by-side alignment (beside = T) for each group, helping the reader to make comparisons between the two datasets. To create a graphics window that allows two graphs on top and one on the bottom, I used the layout() function (see code).

```
> x1 = c(10,20,30,40)
> x2 = c(5,25,35,35)
> my.dat = matrix(c(x1,x2),nrow = 2,byrow = T)
> my.col = gray(seq(0,1.0,length=4))
> layout(matrix(c(1,2,3,3), 2, 2, byrow = TRUE)) # set layout
```

```
> pie(my.dat[1,], col = my.col,radius = 1,
+       main = "Pie Chart 1")
> pie(my.dat[2,], col = my.col,radius = 1,
+       main = "Pie Chart 2")
> leg.txt = c("Pie 1 data","Pie 2 data")
> barplot(my.dat,beside = T, ylim = c(0,50), names = 1:4,
> las = 1,
+         col = gray.colors(2), ylab = "Height (cm)",
+         xlab = "Groups", cex.lab = 1.5, main = "Barplot")
> legend("topleft",leg.txt, fill = gray.colors(2))
> abline(h=0)
```

5.8 Multiple Graphs (Using par() and pairs())

You have already seen several examples where I've managed to place multiple graphs into a single graphics panel (e.g., combining pie charts with barplots in figure 5.9). This is done using the par() function, which controls the making of graphs. You can stack graphs, such as histograms, on top of each other (as seen in figure 5.10) with the par() function, using

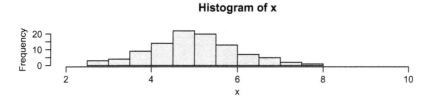

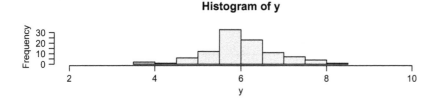

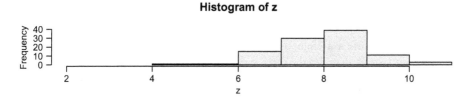

Figure 5.10 A set of three histograms, stacked one above the other, with the help of the par(mfrow = c(3,1)) command (three rows and one column).

a matrix approach to set up the graphics window. In figure 5.10, I've created a graphics window with three rows and one column (> `par(mfrow = c(3,1))`). If you're trying to help your reader see similarities or differences in your data, you should show the graphs on the same scale (e.g., use the > `xlim = c(xmin,xmax)` option in your plot command). I found the ranges of the data using the `floor()` and `ceiling()` functions (see code). In this example, I used the same limits so the range on the `x-axes` are the same and only the data change among the graphs.

```
> # min() gets smallest value; floor() rounds down
> xmin = floor(min(c(x,y,z)))
> # max gets highest value; ceiling() rounds up
> xmax = ceiling(max(c(x,y,z)))
> par(mfrow = c(3,1)) # set graphic window: 3 rows, 1 col
> hist(x,xlim = c(xmin,xmax), las = 1)
> hist(y,xlim = c(xmin,xmax), las = 1)
> hist(z,xlim = c(xmin,xmax), las = 1)
> par(mfrow = c(1,1)) # reset the graphics window
```

Another useful and related graph is a matrix plot, which shows the relationships among several variables simultaneously (figure 5.11). You can

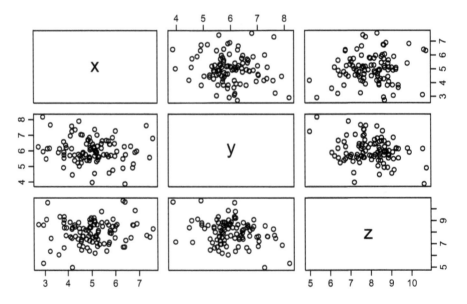

Figure 5.11 A matrix scatterplot using the `pairs()` function. The main diagonal shows no data but instead identifies the three variables (x, y, and z). In the upper-left corners x. The two graphs of data to the right of the x box are graphing x on the y-axis. The two plots directly below the x box have x on the x-axis. As an example, the graph in the middle of the bottom row shows y on the x-axis and z on the y-axis. The data are mirrored on either side of the main diagonal. It's up to you to consider which makes the most sense.

make these using the `pairs ()` function. This is actually really handy to use because it allows you to look for patterns among continuous data and *not* actually do a hypothesis test. This is a type of data mining without actually testing hypotheses.

```
> dat = data.frame(x,y,z)
> pairs(dat)
```

5.9   Problems

1. We're going to use the built-in dataset called "`trees`" that contains height and diameter ("Girth") of black cherry trees. You can view these data simply by typing `trees` in the console and hitting `<enter>`.

    (a) The diameter data are in inches and height values are in feet. Create new variables for diameter and height that are in centimeters and meters called `diam` and `ht`, respectively. Note that you need to use the syntax `trees$diam` and `trees$ht` to pull out the diameter and height data, respectively, from the `trees` dataframe.

    (b) Create a publication-quality graph relating diameter and height of these trees.

    (c) Sort the trees based on height, from lowest to highest and store them in a new dataframe called `trees.sorted`.

    (d) Create a scatterplot of just tree heights against their order in the new dataframe. This should show you how the distribution in height increases.

2. The built-in dataset "`precip`" contains annual rainfall data for cities in the United States in inches. Use it to answer the following problems. Remember to include your code with your answers

    (a) How many cities are represented in the dataset?

    (b) Report which city has the highest rainfall and which has the lowest.

    (c) Examine visually the distribution of the rainfall amounts in these cities using both a boxplot and a histogram.

    (d) Determine whether the highest rainfall observed is technically an outlier.

3. Below are four lines of code that make five samples. Be sure to check out what is in the variable M after you run the provided code.

```
> set.seed(7)
> M = matrix(rnorm(25, mean = 10, sd = 2), ncol = 5)
```

```
> M = as.data.frame(M)
> names(M) = LETTERS[1:5]
```

(a) Create two graphs, side by side. (*Hint:* Use the par() function.) On the left, create a boxplot of these five datasets (A–E). On the right, create a barplot of the sample means. Feel free to make up x- and y-axis labels.

(b) Test whether the samples are normally distributed.

4. Below are the numbers of bacterial counts found in six control and six treatment (trmt) plates.

```
control = 2,3,4,5,6,7
trmt = 5,3,4,5,6,9
```

Provide a side-by-side boxplot of these data. Add labels for the axes, individual treatment levels, and a title.

5. A student conducted a time budget analysis for two gray squirrels on her campus. One was black and the other gray. She recorded the following times in minutes for each activity:

Squirrel	Foraging	Grooming	Playing	Resting
Black	133	78	122	47
Gray	95	22	57	33

(a) Create two pie charts for these data, showing percentages.

(b) Create a side-by-side barplot of these data using minutes.

(c) Discuss the pros and cons of each graph for these data.

Box 5.2. **Optional arguments for visualizations.** The following options, or arguments, can be included to improve the looks of graphs. When provided, they must be separated with commas. An example might look like this: > `plot(x, y, xlim = c(0,10))`.

- `xlim = c(low,high)` and `ylim = c(low,high)`. These control the range of the x and y axes, respectively.
- `type = "n"`. This produces a plot without data. This is used when you want a little more control, such as when adding several sets of data using the functions `points()` or `lines()`.
- `type = "l"`. This produces a line graph without points. Note that it is the letter "el."
- `type = "b"`. This produces both a line and the point symbols.
- `pch = number`. This controls the type of points produced in the `plot()` function. Some of the options for this argument include:

  - `pch = 1`. This is the default and produces small open-circle points.
  - `pch = 16`. This produces small, solid circular points.
  - `pch = "symbol"`. You can replace "symbol" with characters, such as a period (`type = "."`) or a plus sign (`type = "+"`).

- `lty = number`. This produces different line types. The default (1) is a solid line. A dashed line is 2.
- `lwd = number`. This is the line width. The default width is 1.
- `cex = number`. This number controls the size of the symbol, 1 being the default.
- `cex.lab = number`. This number controls the size of the x- and y-axis labels. The default is 1, but I usually like 1.5.
- `main = "Your title"`. Add a title. Using `main = ""` removes a title.
- `xlab = "Your x-label"`. Self explanatory.
- `ylab = "Your y-label"`. Ditto.
- `xpd = F`. Used for barplots when you want the lower limit of the y-axis limit to be a value other than zero.

Additionally, you might consider placing text in your graph using the `text()` function, which might look like this: > `text(x, y, "text")`, where the x and y values are coordinates in the graph where the text will be centered. See the help on the `text()` function for some useful expressions and symbols to add (type `?text` and hit `<enter>` at the console).

CHAPTER SIX

An Overview of Science, Hypothesis Testing, Experimental Design, and Inference

SCIENCE IS BY FAR the best method humans have to understand our universe. You certainly enjoy its products every day and probably use it often to solve your own challenges. For example, you may have had the unpleasant experience of getting into a car, turning the key, and discovering it doesn't start. Your first reaction is undoubtedly to test whether turning the key again will simply do the trick. If it doesn't, you'll likely run through a series of other hypotheses that you think might get the car to start (e.g, shout, curse, kick the bumper, or open the hood and threaten the engine). Alternatively, you might pick up sand and throw it eastward or look skyward for sympathy (or shake a fist). Unfortunately, none of these approaches uses a method that has a scientific basis that actually addresses the problem.

Then again, sometimes unrelated activities coincidentally result in a desired outcome. Dowsing for water often works because you'll run into water if you dig down deep enough just about anywhere. Other times crossing your fingers for "good luck" and baseball players going through a series of odd, superstitious behaviors before batting are acquired because they seemed to work once two years ago. What separates the above methods from science is that *science focuses on using empirical evidence, acquired through rigorous observations and/or experiments, to illuminate the mechanisms that govern how natural systems work.* In this chapter, we'll briefly explore the process of acquiring and interpreting the meaning of scientific information.

6.1 What Do We Mean by the Term *Statistics*?

This book is about biological data analysis. I think of data analysis as being the *process of gathering, summarizing, interpreting, and sharing information.* Statistics is certainly a core discipline in this field. But what do we mean by the term statistics? Here are the two main pieces that encompass statistics for our purposes:

1. Statistics are summary values that *estimate* parameter values. For instance, there is an average value for the height of full grown, adult humans on Earth. There is an exact number, called a parameter, that we will never know. However, we can take a random sample of these humans and determine the average height of this sample. This summary value of a sample is called a statistic.
2. We also think of the procedures we use to formally test hypotheses as statistics or statistical tests (e.g., the common t-test).

In biology our process of getting information usually follows the standards of the scientific method and relies on proper experimental design (see section 6.4). Numbers by themselves do not constitute statistics, and they certainly do not if there is no context. When we attach information to numbers, we can have something with a great deal of meaning.

The simplest statistics are summary values, which are called descriptive statistics. You encounter these all the time, such as the average score on an exam, the fastest time in a race, or even the approval rating of the president's job performance. Even from the esteemed journal *Nature* we can find a comparison of multiple means using t-tests when the data are not normally distributed (fig. 6.1). As we'll learn in chapter 8, this is wildly inappropriate. And because this test relies on a sample's mean, it is important that the mean be a good measure of the center of our data. We can see that, for one of the samples, the mean (where the arrow in figure 6.1 points) doesn't even represent one of the data points. Therefore, this mean value is not just a poor statistic estimating some population parameter in this case. It also is pointless!

In biology, we are often interested in knowing something about a *population*, which usually includes everything in that group (e.g., all wolves on Earth). We generally are restricted to collecting data from a sample, or subset of the population, and use statistical procedures to estimate what we're interested in knowing about the population. For instance, we might be interested in whether two different strains of bacteria have different colony growth rates. We don't test all bacteria but, instead, use samples of colonies that we hope represent the larger population of all bacterial strains. For these types of questions, we are not just summarizing numerical information but are, instead, asking what these samples of data mean in the larger context. This extension of a hypothesis from samples to statements about populations is referred to as inference (see section 6.6 for a further discussion of inference).

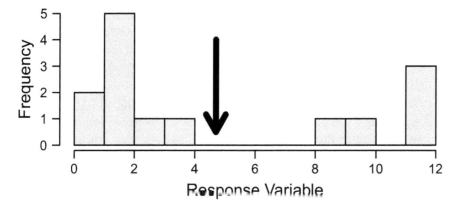

Figure 6.1 A histogram of one of many samples used in a multiple t-test analysis. In this sample there isn't even a data point at the mean (arrow). Not surprisingly, these data are far from normally distributed (data from Zhang 2020).

In this way, we generally collect data on a small subgroup and extrapolate what our results mean to the larger group. Our small group is referred to as a *sample*, while the larger group is called a *population*. Summary values of a sample are called statistics, while summary values of the entire population are called parameters. Therefore, we usually collect samples from populations to get statistics that we use to estimate population parameters. The process of understanding populations from samples using a variety of statistical procedures is referred to as inferential statistics. Much of what we will do in the rest of this book is designed to help you make this leap from summary statistics to inferential statistics.

6.2   How to Ask and Answer Scientific Questions

Here are the basic steps that scientists take when they want to know something about the universe that can't just be looked up. In general, these steps are really hard to accomplish. Scientists usually need extra funding to do this work, and so scientists work to get grants that fund the research. Biological research, in particular, is expensive and time consuming. Therefore, it is quite important that the data be collected carefully and analyzed correctly. Your success in learning to design experiments and analyze the results provides you a valuable and marketable skill. So here are the steps scientists tend to follow, often referred to as the scientific method:

1. Clearly state your question in a way that can be tested. This question is generally stated as a hypothesis.

2. Decide what data are needed to answer the question. These should be just the data you need to answer the question—no more and no less.

3. Hand-draw a graph of what a beautiful answer might look like if your idea is correct. This exercise will always help you clarify your experimental design.

4. Determine the appropriate analysis that would be required to test your hypothesis if you got the cool data you graphed by hand. You can make up data (called dummy data) and try the analysis. Yes, you read that correctly. This way you really understand what you're trying to find out and whether the data you hope to collect will answer your question. If you stop here and present or publish this, you can become famous for a very bad reason.

5. Using the dummy data, you could consider performing a power analysis (see "Power analysis and the number of replicates"). Better is to do this analysis with data from previous experiments or the scientific literature.

6. Design a good way to get those data (experiment or observation). You will likely draw heavily on what others have done before.

7. Do the research and collect the data as carefully as possible.

8. Look at the data you have collected. This is best done with a graph. Test how the data are distributed (normal or not normal). Do the results look like your hand-drawn graph?

9. Test your hypothesis. This is where inferential statistics are used.

10. Share your result with the greater world through a carefully crafted report (e.g., see the journals *Science* or *Nature*) or presentation. Use the style required by your reader (e.g., lab instructor, professor, or journal). It doesn't matter how strange that might seem. Different journals have different styles (as do different faculty members). I once had a student tell me their faculty instructor wanted the methods described in the discussion section of the laboratory report. I asked them if that made sense to them. No, it did not. And I got to say: "And that's how you'll do it, right?"

We need to be sure that we respect the investment made into getting each data point. Time; lab equipment; consumables (e.g., reagents); and the time of faculty members, lab instructors, and assistants are all costly. This leads us to an important rule:

Box 6.1. *The golden rule of data analysis* You should be as careful with your data analysis and developing your visualizations as you are designing and conducting your experiment. Your results are only as valid as the weakest part of your research.

6.3   The Difference Between *Hypothesis* and *Theory*

People often assume that a hypothesis is just a fancy word for a guess in science. This is not correct! A guess is generally some estimate that is made with insufficient evidence or an understanding of the system. For instance, I am an avid tennis player and before a match, players spin a racquet to decide who gets to choose whether or not to serve (e.g., "up" or "down"). The opponent has no information beyond up or down, and we assume that there is an even probability of getting either outcome. The person who gets to choose up or down is making a guess. So what's a hypothesis?

Box 6.2. *A scientific hypothesis* A hypothesis is a well-supported prediction of an outcome that could occur from a scientific investigation.

In general, when we've gotten to the point of developing an experiment, we are testing a well-thought-out hypothesis that has been developed based on previous research experience and/or a deep exploration of the scientific literature. We're never just guessing about what might happen—experiments are too costly! Most often we formally test hypotheses statistically, which usually provides us with some level of confidence, or lack of confidence, in one or more of our hypotheses. And for our purposes in this book, we are generally testing the null hypothesis (H_0). H_0 is one of many potential outcomes and usually agrees with random chance, no pattern, or no relationship. When we do experiments, we often are interested in positive results (e.g., something is happening different from chance), so we test whether our data support H_0 or not. We then report our result and include a p-value. We will develop the technical definition of the p-value very carefully below (section 6.8).

A related term we often hear is *theory*. I once had a mathematician friend of mine state confidently that evolution isn't true—it is "just a theory." He even noted that scientists refer to it as the "theory of evolution." For evolution to be true, must we prove that it is true? Otherwise, isn't it simply a guess? Are scientific theories just guesses? Where's the proof?

The bad news is that we do not prove anything in science. Mathematicians prove stuff and they can do this because, well, they cheat by building proofs that rely on axioms that are *assumed* to be true. In biology, we work to understand nature and continue to challenge our ideas with more evidence. Evidence, however, is always a bit shaky and dependent on many *variables* and sources of *error*. As we work to understand how nature works, we gather more and more evidence that supports or refutes ideas. If the ideas are not supported by data, we may modify our understanding of nature or possibly abandon our ideas. So what's a theory?

Box 6.3. *A scientific theory* A theory is a comprehensive explanation of natural phenomena supported by extensive evidence gathered through observations and/or experiments.

The evidence we gather may increase our confidence that our ideas about nature are correct. But, unfortunately, we only get more and more confident. We never eliminate the possibility that some new evidence might cause us to have to give up on our beloved ideas. Therefore, we cannot be 100% sure an idea is true, which, therefore, prevents us from "proving" anything in science. A slightly controversial corollary to this idea is that, if we can't prove something true in science, we can't disprove it either. If we can't be 100% sure that something is true, then we also can't be absolutely certain that something is false.

What then can we say about our results? We can't say: "our results *prove*" that something is happening. Instead, we say that "our results are *consistent*" with something happening.

So, the words *prove* and *disprove* have limited roles in science, restricted particularly to the area of applied mathematics. Biomathematicians, for instance, do provide proofs that lead us to deep understandings of how biological systems work or can work. Therefore, our inability to prove or disprove hypotheses in science is the result of the inescapable fact that *all results in science are provisional*. For example, we are quite confident the Earth is an oblate spheroid and not flat and that it revolves around the sun. We also have high confidence that species evolve over time and low confidence in spontaneous generation and the medical benefits of homeopathy.

6.4   A Few Experimental Design Principles

Our focus here is to learn to analyze and visualize data correctly. However, to do this correctly we need to, and should know, how best to get data so as to correctly analyze and visualize it. Below are a few important design principles to keep in mind if you're collecting your own data. Most, but not all, observations and experiments will include the items in this list. We should accept, however, that occasionally we can gain great insight about nature without some of these components. Some biological systems, for instance, are just too large and difficult to adequately replicate (e.g., test global atmospheric response to elevated CO_2 or perform experiments on whole-lake ecosystems).

1. **Replication**. This solves the problem that we might have if we only measured one individual who happens to be really unusual. We need, instead, to sample as many individuals as is practical under each treatment level individuals, for instance, under each treatment level so that we can understand the collective behavior of our system. Each individual should be *independent* from other individuals. It is possible, however, to measure a single individual multiple times, but we need to account for this lack of independence. For example, we might measure the mass of an individual over time to determine its growth rate, but we only get one growth rate measurement from this individual and therefore need to make the same type of measurement on other individuals (replicates). A problem might arise if all our study critters are genetic clones within a species. Are these independent replicates? In general, our inference will only extend to this clone, not the species as a whole.

2. **Randomization**. We often need to take individual samples in a study and place them in different treatment groups randomly. If we're sampling out in nature, we need to pick our sites randomly. It's not random, for instance, if we choose research plots that are close to a road or human subjects that happen to be our friends. If we do this, then we can say something only about the groups we are testing, not the larger systems we may actually be wanting to study (see section 6.7).

3. **Factors**. Factors are different treatments that we're investigating and that might affect our study system. For plants, we might be interested in the effect of nitrogen on growth rates. We can have multiple factors that we investigate (e.g., water and temperature).

4. **Levels**. If we have a factor, then we need to have at least two different *levels* of that factor to understand how it influences our study system. In the above example, we might investigate the effect of nitrogen on plant growth rate, and so we have a treatment level with no nitrogen added (*control*) and a level with nitrogen added.

5. **Control**. Controls are tricky sometimes. In the above example, the "no nitrogen" treatment serves as the control. We like to have controls that are factor levels against which we compare our treatments. It is possible, however, not to have a control. If we are interested in testing whether SAT scores for males and females are different, then neither sex would serve as a control group.

6.5   Using **R** to Assign Individuals to Treatments Randomly

When we set up experiments, we often have individuals, or experimental units (some unit for which we will take a measurement), that we need to assign to different treatment groups. Let's imagine that we need to set up a small pilot study to test the effect of a fertilizer on plant size (mass). If we have six plants and two levels of fertilizer (control and fertilized), we should assign three plants randomly to each treatment group.

In such an experiment, we would plant seeds in six identical pots. Each pot should be numbered from 1 to 6 (you should write the number on each pot). We can get our pot numbers like this:

```
> pot = 1:6
```

and we can use the `rep()` function to list the names of the treatments for this:

```
> trmt = rep(c("control","fertilizer"), each = 3)
```

We can now randomize the pots and assign treatment levels to each:

```
> set.seed(11) # so you get the same order as I do (don't do if
> #               doing this for an experiment!)
> pot = sample(pot) # randomize the array pot numbers
> design = data.frame(trmt,pot)
> design
```

```
       trmt pot
1    control   2
2    control   6
3    control   4
4 fertilizer   1
5 fertilizer   5
6 fertilizer   3
```

Undoubtedly a real experiment would be much larger than this, but this skill of randomly assigning individuals to treatments is valuable.

For setting up an experiment, **R**'s function `sample()` is likely one of the most important. You always need some level of randomization of your treatments across your sampling units (the subject you are studying). Alternatively, you may need to choose sample organisms or study plots randomly. To do this, you generally number your subjects or possible study plots, for instance, and then choose or arrange them randomly. The `sample()` function helps you avoid having bias in your experimental design.

6.6   Inference

Once we have designed a study and collected and analyzed our data, we usually want to say something about the natural world as a whole. This statement of the greater meaning of our research is referred to as inference. Our ability to make inferential statements is limited by the data we have collected. If we are interested in endocrinology and have discovered the function of a hormone on metabolic activity in mice, what can we say? We might be tempted to state that our result is important in mammals. We might even go as far as saying that this could be important to humans. Unfortunately, if we've studied this in mice, we can't say this result applies even to related mammal species—and certainly not to humans.

Studies using model systems, such as fruit flies, mice, or yeast, often are done using just one or a few genetic lines where all individuals within a line are genetically identical. In this case, our inference (statement of the greater meaning of our result) cannot go beyond this line of mice. If we use three strains of white lab mice and the result was found clearly in all three, can we say something about this result in white lab mice in general? Possibly, if these three strains were chosen randomly out of all strains of white lab mice. Most likely, however, these mice strains were not chosen randomly but were, instead, strains with which the investigator was familiar or which are simply particularly easy to raise and test under laboratory conditions. Therefore, inferential statements might not go beyond the three specific strains of mice. Think of it this way: if you discover something about three of your friends, are you ready to make an inference about humans?

Power Analysis and the Number of Replicates

If we're designing an experiment, we always have constraints on the size of our experiment. We just discussed the importance of replicates. Why not just have as many as we can afford, have room for, and/or have personnel to do the work? The reason is that we will be making one or more of a variety of mistakes if we do this. We want to be efficient and design the right experiment for the hypothesis we are testing. If we use too many resources, we will not be able to ask other questions. If we use too few resources, then we might not be able to answer our question.

One solution to this problem is for us to conduct a power analysis or a power test. The idea of this is to estimate the number of replicates we need to determine that a factor (e.g., nitrogen) affects our response variable (e.g., plant growth rate). The problem with this determination is that we need to know the answer before we start!

As an example, let's assume that a t-test is the appropriate way to determine if the masses of individual rodents are different between two geographically isolated morphotypes within a species. From previous research, we know that the mean mass of one morphotype is 250 g and that the variability for this species is $s = 25$ g. Given this information, can we estimate how many animals we would have to collect and weigh to detect a significant difference of at least 30 g? The power of a statistical test is defined as $1 - \beta$, where β is the probability of making a Type II error (see section 6.9). Determining power is difficult and depends on sample size, our critical p-value α, the strength of the effect we're studying, and various errors we might make. Often a value of 0.1 is a conservative estimate ($\beta = 0.1$, power $= 1 - \beta = 0.9$). Given this, our power analysis is done as follows:

```
> power.t.test(delta = 30, sd = 25, power = 0.9)
```

```
        Two-sample t test power calculation

              n = 15.61967
          delta = 30
             sd = 25
      sig.level = 0.05
          power = 0.9
    alternative = two.sided

NOTE: n is number in *each* group
```

This suggests that we need at least sixteen replicate individuals (always round up) in each sample group to have our t-test provide us with a significantly different mass for these populations, given that we need the means to be different by 30 g, that the standard deviation of these samples is 25, and that our samples are normally distributed (see chapter 4). This function has more flexibility and other power tests are available.

6.7   How to Set Up a Simple Random Sample for an Experiment

Let's imagine that we are asked to set up an experiment to test the effect of gibberellin on the height of *Brassica rapa* (a fast-growing plant). We would like to employ a proper scientific approach for this experiment so that we can correctly quantify and interpret our results. Let's apply what we learned in the previous section on the principles of experimental design.

1. **Replication**. To test the effect of gibberellin on plant height, we need to be sure that we have several plants. Our laboratory instructor might provide us with ten pots for us to grow replicate plants.

2. **Randomization**. In each pot, we overplant with seeds and, during the first week, we carefully remove all but one plant per pot. At the start of our second week, we are prepared to apply our treatment (spray gibberrellin onto the leaves of the treatment plants). Which plants receive the treatment and which do not? We should do this randomly using the approach described in section 6.4 rather than haphazardly (just choosing them ourselves).

```
> sample(1:10,5) # from the array 1-10, randomly choose 5
```

```
[1]   5   6   4 10   9
```

> If you run this line you should get a different set of plants than I did. These are the pots to treat with gibberellin. The other plants do not receive the treatment.

3. **Factor**. We have just one factor in this experiment: gibberellin.
4. **Levels**. For our gibberellin factor, we have just two levels: with and without gibberellin applied.
5. **Control**. The control plants are those that do not receive gibberellin. In this example, gibberellin is sprayed on the plant leaves at the beginning of week two. Our control should be a spray that doesn't contain gibberellin, such as distilled water. In medical applications you'll hear such treatments as being called a sham or placebo

In addition, we want to be sure to state our hypotheses explicitly. Our hypothesis here might be that giberrilin increases the height of *Brassica rapa* (alternative hypothesis) or that it has no effect (null hypothesis).

6.8 Interpreting Results: What Is the p-value?

A p-value is an important number that comes from conducting many of the statistical tests discussed in this book. It's usually a little tricky to understand at first so let's start slowly. When you conduct one of many statistical tests, you often get a p-value that tells you something about your hypothesis. A simple test might ask whether two samples are statistically different from each other. What we are generally asking is whether the two samples came from the same population. In the previous example, we asked whether gibberellin increases plant height. If the samples are both normally distributed (an assumption of the t-test), then you can perform a t-test (see section 7.6). The test returns a p-value (see figure 6.2). In a t-test, the t-value generally gets larger as the means of the two samples get more and more different (but this also depends on how messy the data are). Larger t-values generally yield smaller p-values and lower our confidence that the two samples came from the same population.

In an effort to make this more intuitive, I will lie and tell you that the p-value is like the probability of the the null hypothesis being true. With this idea in mind, a large p-value, like 0.95, suggests we not reject our null hypothesis, while a small p-value, like 0.001, suggests we should reject our null hypothesis. It's never good to lie so let me formally define the p-value in correct, but more confusing, terms.

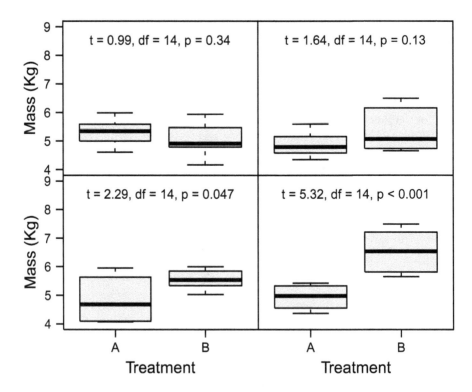

Figure 6.2 Boxplots from four possible studies of the masses of organisms from two different treatments (A or B). In the upper-left panel, it appears that there is no treatment effect (the two samples seem to have come from the same population because p > 0.05. As the samples appear more and more different, our p-values get smaller. At some point, our samples appear so different that we conclude that they did not come from the same population (they are statistically different). We often draw this line at the critical p-value called α, which is often set so that $\alpha = 0.05$.

> **Box 6.4. *The p-value*** The p-value is the probability of getting a test statistic (like *t* returned by a t-test or *F* returned by an analysis of variance [ANOVA] test) that is as extreme or more extreme than the one you observed had the null hypothesis been true.

I assume this definition seems pretty confusing. It's not that bad, however. Statistical tests return test statistics (e.g., a t-value or an F-value). These test statistics have ranges where, at one end, they suggest there's no difference or relationship. At the other end, the test statistic might suggest a big difference between samples or a strong relationship between samples. Generally, when these test statistics are near zero, we *fail to reject the null hypothesis*. Large test statistics reduce our confidence in the null hypothesis. So, for the t-value, for instance, it continues to get larger for samples that are more and more different and, at some point, we reject the null hypothesis that the samples came from the same population (see figure 6.2).

Our definition of a p-value above explains why we consider a p-value that is less than or equal to the critical p-value (α) to be statistically significant. It is important to always set α before we conduct an experiment. If we aren't sure what will happen in our experiment, we generally set $\alpha = 0.05$. Therefore, we would reject the null hypothesis when $p \leq \alpha = 0.05$. Notice, too, that we are dealing with statistical significance; biological significance is quite different.

Where did $\alpha = 0.05$ come from? That's a story involving a famous geneticist named Sir Ronald Fisher. It is an arbitrary value. But it's a widely accepted value. We will use the convention that we fail to reject H_0 when $p > \alpha$. Not everyone will agree with this rule. But it is a simple rule for getting started. Therefore, if $\alpha = 0.05$ and $p = 0.06$, it is wrong to say something like, "The relationship is *almost* significant" or that "there is a positive *trend*." It is even worse to say, "Let's make $\alpha = 0.1$ so it'll be significant!"

6.9   Type I and Type II Errors

As mentioned before, in science we cannot be absolutely certain of anything but instead work to increase our confidence in our understanding. In biology, if we're dealing with data, there always is a chance that we are wrong. This is true even for physicists, at least if they're working with data using a natural system. We accept that there is a probability of making a mistake and have to balance this chance with our desire to interpret our results correctly. Why not just make α really small so that if we reject our H_0, we are really sure the H_0 is wrong? This is a great question!

Imagine we conduct a vaccine trial for COVID-19. We would vaccinate some and not others, and then challenge these brave subjects with the live SARS-CoV-2 virus. If we set our $\alpha = 0.001$, then we would have incredibly strong evidence that the vaccine was effective. But if it worked really well, we risk failing to reject our null hypothesis (the vaccine doesn't work) and people who could be saved with this vaccine would never receive it. Making errors can be dangerous!

Let's consider the situation that we really don't want to make the mistake of incorrectly rejecting a true H_0. For context, let's imagine we're investigating whether human males and females are different in height. We set $\alpha = 0.001$ so that we'll only reject H_0 (no difference in heights) if the heights of males and females are *wildly* different. To conclude that males and

	We Fail to reject the H_o	We reject the H_o
H_o is really true	Correct	Type I Error $= \alpha$ ("false positive")
H_o is really false	Type II Error $= \beta$ ("false negative")	Correct

Figure 6.3 This diagram represents Type I and Type II errors. If we could assume that our null hypothesis (H_0) is true, then, based on our data analysis, we are acting correctly if we fail to reject H_0. If we reject a true H_0, however, then we've made an error (Type I). If we fail to reject a false H_0, however, then we've also made a mistake (Type II). Finally, if we reject a false H_0, then we've acted correctly. Note the probability of making a Type I error is α, while the probability of making a Type II error is β.

females differ in height, we need to see a difference of at least three feet. Then we'd be sure they're different. If males were only 2.5 feet taller than females, it's not enough to be statistically significant. You probably can tell that we've made a mistake (see figure 6.3). What we've done is to avoid making a Type I error by making α (the probability of making a Type I error) very small. But, unfortunately, under this scenario, we've failed to conclude a difference in height between males and females where there really is a difference (2.5 feet on average). We've committed what's called a Type II error.

This becomes more problematic when the results matter, such as in the case of a new drug that could save or extend lives. If the life expectancy of patients increased because of our drug, what length of time might be considered significant? We are likely to think one week might be great. If we set α to a small value, we would not likely detect such a difference. We would be really rigorous in our result by wanting to be *certain* that our drug extends life by a long time. If our drug does extend life but, perhaps, not by much, or the result is pretty variable, we *fail to reject* our false H_0 (Type II error; see figure 6.3). We might end up not recommending that a drug that works be administered. Are you comfortable with that?

To avoid this mistake, we might set $\alpha = 0.2$. This is great because the test is really sensitive (it's easy to reject our H_0). Is this a problem? This error might be acceptable because maybe the drug is inexpensive and has a placebo effect, so even if it doesn't work, no one is hurt (see figure 6.3). We often must consider which of these errors is more acceptable.

In this chapter, we've discussed a variety of ways to conduct sound science and assess information. As you can see, this process is quite challenging to do correctly. And its complexity ensures that people, not just statistical programs, play a critical role in order to interpret results correctly.

6.10   Problems

1. You take your first test in a biostatistics class and the professor reports that the average (mean) was 75%

   (a) The reported mean is a model representing a simplification. a single number represents all grades earned by students in the class. What assumptions has the professor made in order to report the mean *correctly*?
   (b) Is the reported mean, in this example, a statistic or a parameter?
   (c) What inference can be made regarding this result?

2. You perform a statistical test and get a p-value of 0.06.

   (a) Under what conditions would we consider this to be statistically significant?
   (b) Under what conditions would we consider this not to be statistically significant?
   (c) Provide a biological example where a p-value $= 0.06$ might not be statistically significant but is, instead, biologically significant.
   (d) Assuming the researcher set $\alpha = 0.05$ prior to conducting the experiment, describe the *type* of error made by concluding that $p = 0.06$ is close enough to being significant that H_0 is rejected.

3. You walk into your bedroom and flip on the light switch. The light does not go on. Assuming this is a problem you want to fix, describe the steps that you would take to remedy this using the scientific method.

4. You plan to investigate how the drug enalapril affects blood pressure in patients. You advertise in a local newspaper to pay anyone over 50 years of age with hypertension $100 to to join the study. From the first thirty to sign up, you create a randomized controlled trial (RCT) with three groups, each with ten people. The groups receive either a placebo, 2.5 mg, or 5 mg daily dosages. You measure their blood pressure before the trial begins and after one month.

   (a) Provide a hand-drawn graph of what the results will probably look like. Include a reasonable scale and units.
   (b) Describe what the control is (or should be) in this experiment.

(c) Provide code that would randomly place ten individuals into each study group.

(d) What is the population to which your results would apply?

(e) Describe the conditions under which you would want $\alpha = 0.01$ for this trial.

(f) Describe the conditions under which you would want $\alpha = 0.2$ for this trial.

5. Three student researchers are interested in how a particular plant species responds to elevated atmospheric CO_2. They clonally propagate cottonwood saplings (*Populus deltoides*) to reduce the variability among individuals and that therefore maximizes their ability to attribute any response they see to the CO_2 treatment effect. What inference can they make about plant response to changes in CO_2?

6. Below is a quote from a *New York Times* (September 12, 2012) article titled "How Testosterone May Alter the Brain After Exercise."

> A new study published last month in the *Proceedings of the National Academy of Sciences* [found] that male sex hormones surge in the brain after exercise and could be helping to remodel the mind. The research was conducted on young, healthy and exclusively male rats—but scientists believe it applies to female rats, too, as well as other mammals, including humans.

(a) Describe the population that these results apply to.

(b) Describe why the results can or can't apply to female rats.

(c) Describe why the results can or can't apply to humans.

7. In section 6.7 we discussed growing ten plants in ten separate pots, with five plants receiving the gibberellin treatment and five serving as our controls. To save money, the researcher decides to grow the ten plants in two pots, five per pot. They spray the plants in one pot with gibberellin after one week and spray the other plants in the other pot with an equal amount of distilled water (the control plants).

(a) Describe what problem or problems could arise from this experimental protocol (check out the concept of pseudoreplication, first described by Hurlbert (1984).

(b) Why would it be important for the control plants to receive a treatment spray of distilled water at the same time gibberellin is applied to the treatment plants?

CHAPTER SEVEN

Hypothesis Tests: Using One- and Two-Sample Tests

THIS CHAPTER TACKLES HYPOTHESIS TESTING when we have one sample and a test value or two samples that we want to compare against each other. Choosing which test to use and how to implement them correctly is tricky. I've worked to group tests together based on the structure of your data. However, which test you use may be influenced by experience and by suggestions of your laboratory instructor or research mentor. If you are not sure what to do, keep in mind you've learned a lot so far and should be able to ask a qualified person for directions and implement the solution with one of the techniques introduced in this and the next few chapters. The tests presented here form a starting point for you. No matter your questions, however, the answers can be obtained from **R**, so you're definitely on your way to solving a wide variety of problems in biology.

We begin with one-sample tests. The data may or may not be normally distributed. In addition, the questions we ask might be one- or two-tailed, meaning that we want to know whether there's a simple difference or whether one sample is, on average, greater or less than another value or sample. If there is direction in our question, then we will be conducting a one-tailed test. If our question does not have directionality, then we will be using a two-tailed test.

7.1 One-Sample, Two-Tailed Test with Normally Distributed Data

Below are combined SAT scores for twenty randomly chosen students at your school. We'll assume that your combined SAT score is 1120. We'd like to know whether your SAT score is statistically different from this random group. In other words, is the sample of student SAT scores and your score all drawn from the same population of SAT scores or not?

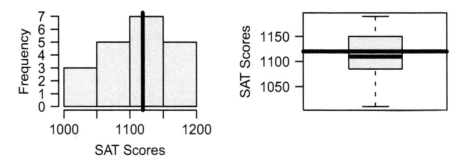

Figure 7.1 Visualizations of the SAT scores for a sample of university students using a histogram (left) and a boxplot (right). The thick lines in both graphs represent your SAT score. The histogram *suggests* the data may be normally distributed. In the boxplot, we see that the median is close to the middle of the data, also consistent with normally distributed data. We also note that your score of 1120 is found near the middle of the distribution of SAT scores from the random student sample.

```
> SAT.scores = c(1130, 1090, 1190, 1110, 1160, 1120, 1160, 1110,
+               1080, 1160, 1050, 1120, 1030, 1010, 1080, 1090,
+               1170, 1110, 1090, 1140)
> your.SAT = 1120
```

First, we should visualize the data, including the sample and your SAT score. This can be done using a histogram and/or a boxplot (see figure 7.1 and Box 7.1). It doesn't have to be fancy. The data seem like they are normally distributed. We note that your score appears to fall near the middle of the sample distribution and, therefore, we will ask whether your SAT score is different from the sample SAT scores. This question does not have directionality and thus is a two-tailed test. (the test value, which is your score, might be in either tail of the distribution).

Next, we need to test whether the sample is normally distributed. As we learned in chapter 4 (section 4.4) we can use the Shapiro-Wilk test. Here is the test return for the p-value only, for brevity.

```
> shapiro.test(SAT.scores)$p.value
```

```
[1] 0.7522183
```

Recall that the null hypothesis for the normality test is that the data are normally distributed. Because our p-value of 0.75 is greater than 0.05, we conclude that our data meet the assumption of normality for a t-test.

Box 7.1. Boxplots versus barplots. Many researchers, and the journals that they publish in, prefer presenting data using barplots with error bars (showing variability in the samples) over boxplots. This is unfortunate for three reasons:

1. Conion doesn't make it right.
2. Boxplots show more information about the distribution of data than do barplots with error bars.
3. Barplots with error bars assume data are normally distributed. They often aren't, and the ability to know is hidden from the reader.

It is always appropriate to show boxplots and sometimes appropriate to use barplots and error bars. So which one should you use? Whichever the person who's evaluating your work tells you to use, of course! If you must use barplots with error bars, here's one way to do it in **R**. The generally preferred error bar, is a 95% confidence interval (shown below). Other error bars include the standard deviation, standard error of the mean, and variance.

```
> par(mfrow = c(1,2))
> library(plotrix) # install this package, if necessary
> boxplot(SAT.scores, ylab = "SAT Scores", cex.lab = 1.5,
+         ylim = c(1000,1200), las = 1)
> M = mean(SAT.scores)
> s = sd(SAT.scores)
> SEM = s/sqrt(length(SAT.scores))
> CI95 = qt(0.975, df = length(SAT.scores) - 1)*SEM
> a = barplot(M, ylim = c(1000,1200), xpd = F, cex.lab = 1.5,
+             ylab = "Mean SAT Scores", las = 1)
> abline(h=1000)
> plotCI(a, M, CI95, pch = NA, add = T, lwd = 2)
```

Before we do this, it is necessary (and helpful) to declare our hypotheses formally. These will include one *null hypothesis* (H$_0$) and at least one *alternative hypothesis* (H$_A$). The alternative hypothesis mirrors exactly how our question is asked. In this case, we're asking, Is the mean of the sample of SAT scores different from the test value (μ = 1120), given the level of variability in the sample data? Therefore, H$_A$ (alternative hypothesis) is that the sample and test value are not equal. The null hypothesis, which is what **R** will actually test, includes all other options. Here are these two hypotheses:

H$_0$: $\bar{x} = \mu$
H$_A$: $\bar{x} \neq \mu$

R requires that we send the t.test() function the sample SAT.scores and the test value (in **R** μ is spelled as "mu"). By default, the t.test() function assumes $\mu = 0$ and that we're requesting a two-tailed test. In our case, mu = 1120. We are now ready to conduct our one-sample, two-tailed t-test.

```
> t.test(SAT.scores, mu = your.SAT) # mu value tested against
> sample
```

```
        One Sample t-test

data:  SAT.scores
t = -0.95119, df = 19, p-value = 0.3535
alternative hypothesis: true mean is not equal to 1120
95 percent confidence interval:
 1087.996 1132.004
sample estimates:
mean of x
    1110
```

This is the output from the t-test. For us, the important part is the statistical test result, which starts with t = -0.951. We've tested the data using a two-tailed approach, meaning that we've asked whether the sample is either less than or greater than our test value (your score). Our result suggests that the sample of SAT scores from other students is not statistically different from your SAT score. A proper results statement would say this and end, parenthetically, with this: (t = −0.951, df = 19, p = 0.354). So we've concluded that the sample of SAT scores and your SAT score are not different (they come from the same population).

You might wonder why the t statistic is negative. This actually provides us some interesting information about the relationship between the sample mean and the test value. The t statistic is calculated as follows:

$$t = \frac{\bar{x} - \mu}{SEM} \tag{7.1}$$

where $\bar{x}$ is the sample mean, the test value is μ, and *SEM* is the standard error of the mean of just the sample (the test value is not included). Because

$\bar{x} < \mu$ the numerator in equation 7.1 is negative. Let's verify that **R** correctly calculated the t value:

```
> SEM = sd(SAT.scores)/sqrt(20)
> (mean(SAT.scores) - your.SAT)/SEM
```

```
[1] -0.9511897
```

Therefore, if the t-test returns a negative t-value, then our sample is, on average, smaller than our test value. We also see that, in this case, although the mean is smaller than mu, it is *not* significantly different.

7.2   One-Sample, One-Tailed Test with Normally Distributed Data

I asked students how long they can hold their breath. I'm pretty competitive and so I like saying they can't hold their breath as long as I can. Students then hold their breath and try to beat me. Their times in seconds are below:

```
> breath = c(33, 79, 41, 41, 25, 51, 50, 46, 53, 61)
```

The first thing we can do is a quick visualization of the student sample times and the test value (my time of 65 seconds). For this, I'm will produce quick and simple graphs (see figure 7.2)

From figure 7.2, we see that students, in general, were not able to hold their breaths as long as I could. Note that we have not yet tested a hypothesis. The original challenge suggested that I could hold my breath *longer* than the students. The proper way to state this question, however, is with respect to the sample, not the test value. The question is, therefore, Is the sample statistically less than the test value? This becomes our alternative hypothesis (H_A).

Before we do this test, we need to determine whether we'll be conducting

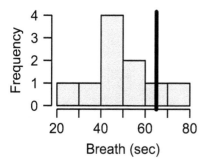 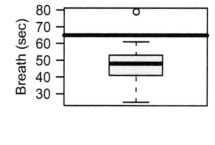

Figure 7.2 The distribution of the amount of time that students could hold their breath. The test value (65 seconds) is the thick line in both graphs.

a parametric test (e.g., t-test) or some other test (we'll talk about those soon). So let's test the sample for normality:

```
> shapiro.test(breath)
```

```
        Shapiro-Wilk normality test

data:   breath
W = 0.96315, p-value = 0.8211
```

Because the data are statistically normally distributed ($p > \alpha = 0.05$), we know that we can use the t-test, which relies on the difference between the mean and the test value. Therefore, our hypotheses look like this:

H_0: $\bar{x} \geq \mu$
H_A: $\bar{x} < \mu$

We are now ready to test the hypothesis. We write the **R** code as follows:

```
> t.test(breath, mu = 65, alt = "l")
```

```
        One Sample t-test

data:   breath
t = -3.5848, df = 9, p-value = 0.002943
alternative hypothesis: true mean is less than 65
95 percent confidence interval:
      -Inf 56.69307
sample estimates:
mean of x
      48
```

Note that the `alt` argument is set equal to the letter "l," not the number "1." If our question had been: Can students hold their breaths longer than I can? then we would simply use the argument "`alt = g`." From this analysis, we conclude that the students, on average, did not hold their breath as long as I did ($t = -3.585$, df = 9, $p < 0.01$).

7.3   One-Sample Tests with Non-Normally Distributed Data

We can conduct a similar test that is non-parametric if our sample fails the normality test (the Shapiro-Wilk test returns a p-value $\leq \alpha$). The function `Wilcox.test()` can be used. Perhaps we wish to test whether a group of locals and their awesome jumping frogs (called the "jumpers") can out jump our "Celebrated Jumping Frog of Calaveras County" (see the 1867 story by Mark Twain). I'm going to claim that this frog could jump 137 cm in one leap. Note that this is a directional test.

Assuming we have a sample of distances from the local frogs (in cm) in a variable called `jumpers`, we should first graph them (graph not shown):

```
> jumpers = c(48, 118, 136, 134, 129, 123, 119, 87, 106, 119)
> par(mfrow = c(1,2))
> hist(jumpers)
> abline(v = 137, lwd = 3)
> boxplot(jumpers)
> abline(h=137, lwd = 3)
```

From these graphs, we grow suspicious that these data may not be normally distributed. Let's test this:

```
> shapiro.test(jumpers)$p.value
```

```
[1] 0.01600322
```

From this result, we conclude that the sample of distances is not normally distributed.

We now are ready to test whether this sample of frogs jumps farther than the celebrated frog. Note that the data suggest the celebrated frog jumps farther but that our hypothesis test is whether the sample jumps farther. Again, the hypothesis is always in terms of the sample, so our hypotheses can be formally stated as follows:

H_0: $\bar{x} \leq \mu$
H_A: $\bar{x} > \mu$

and our final statistical test should look like this:

```
> the.frog = 137 # distance in centimeters
> wilcox.test(jumpers, mu = the.frog, alt = "g")
```

```
        Wilcoxon signed rank test with continuity correction

data:  jumpers
V = 0, p-value = 0.9979
alternative hypothesis: true location is greater than 137
```

From this analysis, we conclude that the sample of frogs did not jump as far as the celebrated frog of Calaveras County (V = 0, df = 9, p = 0.998).

7.4   Paired Data That Are Normally Distributed

Sometimes we're interested in how our study system changes over time. For instance, we might have a measurement of an organism before and after some sort of treatment. These types of measurements would not be *independent* measures but are, instead, called repeated measures or paired measures. As mentioned above, I have been asking students for years to determine how long they can hold their breath. It's been my observation that students, in general, are able to hold their breath longer the second time they try.

Below is a subset of data from a class from a few years ago. We're going to want to answer the question, Can students hold their breaths longer during the second trial? Note that this is a directional question—I'm not asking whether the length of times are simply different. Here are the data with values in seconds:

```
> first = c(106.70, 72.00, 58.80, 48.00, 53.53, 35.93, 39.91,
+           31.00, 45.85, 78.50)
> second = c(129.00, 101.00, 64.20, 58.00, 64.78, 50.92, 48.50,
+            42.09, 70.00, 124.60)
```

Note that it is important that the measurements are ordered by students so that each student has the same index value in both datasets.

We should first graph these data. We might first be tempted to create a boxplot of these two samples side-by-side for comparison (see figure 7.3). We can see that it seems students are able to hold their breaths, on average, longer. However, this is an incorrect way to graph these data! Graphing the data this way implies that the samples are *independent*. They are not! Each student is in both the first and second attempts! It is okay to measure individual experimental units more than once, like these students, but we can't treat these measurements as independent. What we need is a single value of change for each student. We can calculate this difference as follows:

```
> breath.diff = second - first # create a new variable
```

We should graph these data. In figure 7.3, I've graphed these data two ways. Be sure to convince yourself that one way is appropriate and the other is quite inappropriate for these data. Another good method to visualize these

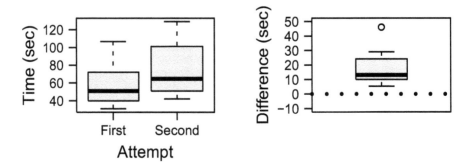

Figure 7.3   On the left are boxplots for the first and second attempts by students to hold their breaths (time in seconds). We see that, collectively, times seem to increase. However, this is an incorrect representation of the length of time students can hold their breath. The graph on the right shows the additional time in seconds that students could hold their breath during their second attempt compared to their first attempt. This is the correct way to show these data. The reference line at zero represents no change in time.

data is to use a bump chart that shows the change in time for each student (see figure 5.8). This is left as an exercise.

Now that we have this difference value for each student, we can test whether that difference improved. For it to improve, we would expect the variable `breath.diff` to be statistically *greater* than zero. We first need to test whether the `breath.diff` data are normally distributed.

```
> shapiro.test(breath.diff) # test this single sample for
> normality
```

```
         Shapiro-Wilk normality test

data:  breath.diff
W = 0.86885, p-value = 0.09692
```

We conclude that the data are normally distributed, so we may proceed with our parametric test. This, again, becomes a one-sample t-test. The original question asked whether the length of times students could hold their breath increased, which is the same as asking whether the difference (second, longer attempt minus the first, shorter attempt) is greater, on average, than zero. We can write our hypotheses like this:

$H_0: \bar{x} \leq 0$
$H_A: \bar{x} > 0$

We can now perform the test as follows:

```
> t.test(breath.diff, alt = "g")
```

```
         One Sample t-test

data:  breath.diff
t = 4.6705, df = 9, p-value = 0.0005839
alternative hypothesis: true mean is greater than 0
95 percent confidence interval:
 11.10951      Inf
sample estimates:
mean of x
   18.287
```

From this analysis, we conclude that the students did significantly increase the length of time they could hold their breath (t = 4.67, df = 9, p < 0.001). The appropriate visualization for this result is the right graph in figure 7.3. One last note: it is equally valid to test the separate times using the `t.test()` function, but we need to tell **R** that these data are paired:

```
> t.test(second, first, paired = T, alt = "g") # T for TRUE
```

The argument `alt = "g"` is read in these tests as the first argument (`second`) is *greater* than the second argument (`first`). The problem with this approach, apart from being confusing, is that we still need to determine whether the data are normally distributed to use this test appropriately or know that we need to use the non-parametric Wilcoxon test (below). So it is best simply to create a new variable for the differences and correctly use that new sample. Always check the answer you get from your statistical test with your graph of the data: they should be in agreement.

7.5   Paired Data That Are Not Normally Distributed

If the data are not normally distributed we, instead, need to substitute the function `wilcox.test()` for the `t.test()` function. The arguments sent to these functions are the same.

The LDL cholesterol levels of individuals were measured for ten human subjects. These individuals were then "treated" by restricting them to meals served only at an unnamed fast-food restaurant for thirty days. We will ask whether there was a change in LDL for these individuals. Note that this is not a directional test. Here are the data:

```
> before = c(75, 53, 78, 89, 96, 86, 69, 87, 73, 87)
> after = c(124, 103, 126, 193, 181, 122, 120, 197, 127, 146)
```

Because we are interested only in the changes, we test whether the differences (`after-before`) are normally distributed:

```
> LDL.diff = after-before
> shapiro.test(LDL.diff)$p.value
```

```
[1] 0.02916288
```

The p-value for the Shapiro-Wilk test is less than 0.05, so we reject the null hypothesis that the data are normally distributed. Therefore, we proceed with the non-parametric Wilcoxon test. We can write down our hypotheses, noting that we are not asking about the mean but are, instead, asking about the median (represented by an x with a tilde over it: $\tilde{x}$).

H_0: $\tilde{x} = 0$
H_A: $\tilde{x} \neq 0$

We can now perform our test, which, being a two-tailed test with a test value of zero, is written like this:

```
> wilcox.test(LDL.diff)
```

```
        Wilcoxon signed rank exact test

data:  LDL.diff
V = 55, p-value = 0.001953
alternative hypothesis: true location is not equal to 0
```

Note that there is a significant difference in LDL levels before and after the treatment. In particular, the LDL of our sample of individuals increased significantly after one month of eating at a fast-food restaurant (V = 55, p = 0.002). The graphing of these data would follow the same format that we saw for the paired t-test. Note that no individuals were actually hurt to get these data.

7.6   Tests with Two Independent Samples

In the previous section, we explored in detail examples where there was only one sample. We saw data that were either normally or non-normally distributed and asked questions that were either directional (one-tailed) or nondirectional (two-tailed). We're going to extend this briefly to two samples, recognizing that the differences in approaches are not that different when we have two independent samples.

Samples Are Normally Distributed

For this example, we're going to look at an experiment that tested whether adding fertilizer to plants causes them to grow taller. Note the directionality in the test. Below are the data:

```
> cont = c(64.7, 86.6, 67.1, 62.5, 75.1, 83.8,
+ 71.7, 83.4, 90.3, 82.7) # control plants
> fert = c(110.3, 130.4, 114.0, 135.7, 129.9,
+ 98.2, 109.4, 131.4, 127.9, 125.7)
# fertilized treatment plants
```

We always begin the analysis by visualizing our data. A side-by-side box-plot is an appropriate approach (figure 7.4).

```
> boxplot(cont, fert, names = c("Control","Fertilizer"),
+         xlab = "Treatment", ylab = "Plant Height (cm)",
+         ylim = c(60,140), cex.lab = 1.5, las = 1)
```

The graph in Figure 7.4 seems to suggest that fertilizer increases plant height, as we suspected. We also see that the two samples might be normally distributed. Let's test these samples for normality. Because these two samples are independent, we need to test them individually. Here are just the p-values from the normality test:

```
> shapiro.test(cont)$p.value # normal?
```

```
[1] 0.3725881
```

```
> shapiro.test(fert)$p.value # normal?
```

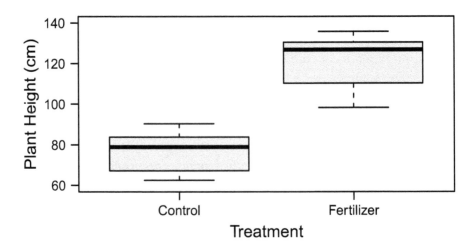

Figure 7.4 A comparison of the heights of plants with fertilizer and without (control). The fertilized plants are significantly taller than the controls (t = −8.884, df = 18, p < 0.001). Do the data in the graph support this statistical finding?

```
[1] 0.1866711
```

The samples are statistically normally distributed, so we may be able to proceed with a standard t-test. However, the t-test also assumes that the variances are equal. Let's test that, too, using the var.test() variance test.

```
> var.test(cont,fert)
```

```
        F test to compare two variances

data:  cont and fert
F = 0.63802, num df = 9, denom df = 9, p-value =
0.5137
alternative hypothesis: true ratio of variances is not equal
        to 1
95 percent confidence interval:
 0.1584741 2.5686485
sample estimates:
ratio of variances
        0.6380159
```

The null hypothesis for this test is that the variances are not statistically different. The resulting p-value (p = 0.51) suggests that the variances aren't different, so we may proceed with a t-test instead of the Welch test (the default test, done if variances are unequal). We now know that we can use a parametric t-test to test our hypothesis about whether fertilizer increases this plant's height. To do this, we write down our hypotheses as we've done before:

$H_0: \overline{fert} \leq \overline{cont}$
$H_A: \overline{fert} > \overline{cont}$

where $\overline{fert}$ represents the mean for the `fert` treatment. Our prediction is the alternative hypothesis, while the null hypothesis is everything else (including no change in plant height).

Finally, it's time to conduct the statistical test. Remember to include the argument `var.equal = TRUE` to our function call.

```
> t.test(cont, fert, alt = "l",var.equal = TRUE)
```

```
        Two Sample t-test

data:   cont and fert
t = -8.884, df = 18, p-value = 2.67e-08
alternative hypothesis: true difference in means is less than 0
95 percent confidence interval:
      -Inf -35.81405
sample estimates:
mean of x mean of y
    76.79    121.29
```

This is confusing. Why did I write `alt = "l"`? The way to read the arguments is as follows: the alternative hypothesis is that the first argument (`cont`) is less than the second argument (`fert`). That is the same as the our stating that the `fert` heights are greater than the `cont` heights. Therefore, we could have written the `t.test()` as follows:

```
> t.test(fert, cont, alt = "g", var.equal = TRUE)
```

It is always important to be *very* careful about how we ask and answer these seemingly simple statistical questions! From this result, we conclude that fertilizer significantly increased the height of plants ($t = -8.884$, df = 18, $p < 0.001$, see figure 7.4).

Samples Are Not Normally Distributed

A dairy cow can produce a lot of milk, about 10,000 kg in a year. Below are data for two samples of cows treated with different levels of antibiotics to help maintain the health of the cows. Do cows produce different amounts of milk depending on these two treatments?

```
> A = c(5800, 5000, 6500, 5000, 5200, 7900, 5300, 5900, 7600,
> 5400)
> B = c(7100, 8600, 8900, 7500, 7300, 7100, 7400, 8300, 7000,
> 7000)
```

We should first visualize these data. A boxplot would be appropriate (see figure 7.5):

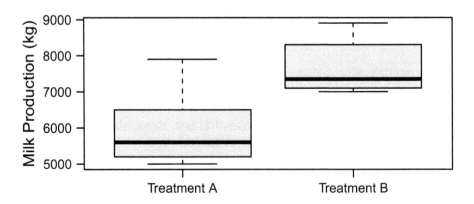

Figure 7.5 Milk production for two samples of cows.

```
> boxplot(A,B, names = c("Treatment A", "Treatment B") las = 1,
+          ylab = "Milk Production (kg)", cex.lab = 1.5)
```

Both sets of data appear to be skewed to the right. We need to test these samples for normality:

```
> shapiro.test(A)$p.value # normal?
```

```
[1] 0.04226108
```

```
> shapiro.test(B)$p.value # normal?
```

```
[1] 0.02425519
```

The samples are not statistically normally distributed. If we log-transform these data, then sample B remains non-normally distributed. Given this, we will proceed with the non-parametric test using the original data. We can write down our hypotheses in terms of sample median and recall that our question was whether these samples were simply different (no directionality):

$H_0: \widetilde{A} = \widetilde{B}$
$H_A: \widetilde{A} \neq \widetilde{B}$

With these hypotheses, we can now perform the non-parametric test. We need to send the `wilcox.test()` function only the two samples. By default, the function will assume that there is no directionality in the hypothesis test (no need to include an "`alt = `" argument) and that we're just looking for a difference between the medians (so there is no expected difference between the sample sizes, so no "`mu = `" argument).

```
> wilcox.test(A,B)
```

```
          Wilcoxon rank sum test with continuity correction

data:   A and B
W = 14, p-value = 0.007219
alternative hypothesis: true location shift is not equal to 0
```

From this result, we conclude that milk production by the two samples of cows treated with different antibiotics were statistically different. In particular, cows receiving antibiotics in treatment B outproduced cows receiving treatment A (W = 14, p = 0.007219, see figure 7.5).

7.7   Problems

1. Below are the numbers of adult smaller tea tortrix moths (*Adoxophyes honmai*) counted in different light traps (Nelson, Bjornstad, and Yamanaka, 2013).

    1916, 1563, 1436, 6035, 3833, 5031, 13326, 3130, 6020, 1889

   (a) Enter the data into a variable called moths and test whether the data are normally distributed. If they are not normally distributed, test whether they are after log-transforming them (use the log() function).
   (b) If we collect data from a single light trap this year and count 10,000 adults, would this be considered a statistically higher number of moths? Provide the written hypotheses for this test.
   (c) Perform the correct statistical test for question 1(b).
   (d) Write a complete results statement, including the proper statistical output to support your claim.

2. Create a bump chart for the data on students holding their breath twice.
3. The order of leaves along the stem of plants is referred to as phyllotaxis. A researcher is interested in whether the fifth order leaves (lf5) differ in leaf area from the first order leaves (lf1). The leaf areas, in cm^2, for six plants are shown below.

$$lf1 = 27, 26, 20, 24, 28, 29$$
$$lf5 = 30, 34, 28, 35, 42, 40$$

   (a) Note that the leaves are paired for the six plants. Are the data normally distributed?
   (b) Provide the hypotheses for this test.
   (c) Complete the correct statistical test to answer the question.
   (d) Create an appropriate visualization of your result.
   (e) Provide a complete results statement.

4. Two unnamed universities (U1 and U2) compete in a championship basketball game. The height of the twelve players on each team are listed below, in inches. Sports pundits say height matters in the game. You want to know if there is enough evidence to suggest that one university has an advantage over another with respect to height.

U1 = 81.0, 80.1, 86.1, 78.9, 86.8, 84.6, 79.3, 84.0, 95.4, 70.3, 86.8, 78.1

U2 = 94.4, 76.7, 70.0, 88.8, 73.7, 86.3, 85.7, 74.0, 79.5, 75.9, 68.1, 75.9

(a) Are the data normally distributed?
(b) Provide the hypotheses for this test.
(c) Complete the correct statistical test to solve the problem.
(d) Create an appropriate visualization of your result.
(e) Provide a complete results statement.

5. Below are combined SAT scores for a random sample of ten undergraduate students from a randomly selected college and another sample of undergraduates from a university center. You want to know if the SAT scores are different between these institutions, based on these sample.

Institution	SAT Scores
College	1330, 1320, 1350, 1370, 1390, 1470, 1340, 1470, 1450, 1360
University	1190, 1160, 1140, 1390, 1360, 1320, 1150, 1240, 1380, 1180

(a) Are the data normally distributed?
(b) Provide the hypotheses for this test.
(c) Complete the correct statistical test to solve the problem.
(d) Create an appropriate visualization of your result.
(e) Provide a complete results statement.

CHAPTER EIGHT

Hypothesis Tests: Differences Among Multiple Samples

WE OFTEN ARE INTERESTED in testing the effect of a factor over a variety of levels. For instance, if we're testing the effect of nutrition on an organism's growth rate, we might have a treatment without the addition ("control"), as well as levels with medium and high additions. When we have more than two sample groups, it's not statistically correct to do just multiple t-tests. Why? The bottom line is that conducting multiple tests, like many t-tests, increases our chance of making a Type I error (see the example in chapter 6). You may recall from section 6.9 that this happens when we erroneously reject a true H_0. The use of a single test, like an analysis of variance (ANOVA), reduces our likelihood of making this mistake.

There are many types of ANOVA. We will assume only one type: a fixed effects model. In this design, you collect continuous data measurements (real numbers) for your response variables, and the different levels chosen for factors are exactly the ones you want to test. This makes the factors fixed. Alternatively, you may have a large number of levels that you could choose from and so you randomly pick those. An example of this design might be that you want to test how new football helmets help reduce concussive injuries, and you have many teams to choose from. Therefore, you choose teams at random. You'll want to pursue information on how to conducted this type of random effects or mixed effects model, which is not discussed here. Finally, there also are other designs, such as repeated measures, nested designs, and blocking effects, which are beyond the scope of this chapter.

8.1   One-Way ANOVA

The ANOVA test is generally used when we have more than two samples that are each normally distributed. In addition, it can be used when you have just two samples and your conclusion will be exactly the same as a t-test. In general, with this test, we are asking whether the samples come from the same population or not, using the variability within and between samples (see box 8.1).

Box 8.1. *Analysis of variance or analysis of means?* This test is quite different from a t-test, which you may recall tests the difference between two sample means. An ANOVA, however, does not test the means. Instead, it tests a ratio of variances. In the left graph below are three samples, each with large *within*-sample variances. There isn't that much *between*-sample variance. The opposite is true for the data in the graph on the right. An ANOVA tests the ratio of the between-sample variance to the within-sample variance. Therefore, the data in the right graph will yield a much smaller p-value than the data in the left graph.

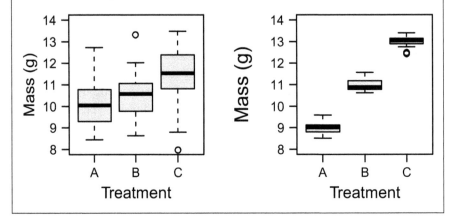

Let's assume we have data on the mass of young fish that were grown using three different food supplemental rates. When we have more than two samples and the data are normally distributed, then we are able to use an ANOVA to test whether there was an effect of food treatment on fish growth, measured as the mass of fish grown over some length of time (here, one month). One challenging part of this test is getting the data into the proper format. To do this, I have simply entered the mass of each fish grown under each treatment level into its own variable in an RStudio script file. Then it's simple enough to gather the data into a dataframe and stack them (we discussed stacking data back in chapter 3).

```
> Low = c(52.3, 48.0, 42.3, 50.8, 53.3, 45.1)
> Med = c(50.4, 53.8, 53.4, 58.1, 56.7, 61.2)
> High = c(66.3, 59.9, 57.1, 61.3, 58.3, 55.4)
> fish.dat = data.frame(Low,Med,High)
> fish.stacked = stack(fish.dat)
```

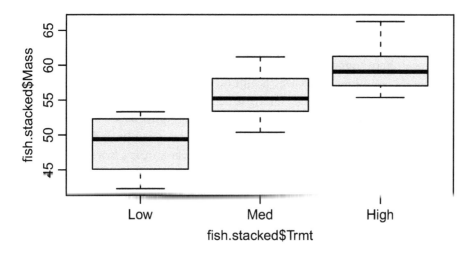

Figure 8.1 A simple boxplot of fish growth under three feeding rates.

The default variable names after stacking are "values" and "ind." We can change these using the names () function:

```
> names(fish.stacked) = c("Mass","Trmt")
```

In this stacked format, we can create a quick graph of the data using the formula method (y ~ x) to see the distributions of the samples (figure 8.1).

```
> boxplot(fish.stacked$Mass ~ fish.stacked$Trmt)
```

These data appear to be normally distributed, but we must, of course, test this formally. It is important not to test all the data at once to see if they are normally distributed collectively. Instead, we must test the individual samples separately. We can do this in a variety of ways. You can use your skill with either the which() or subset() functions to extract just the data values for each level (low, medium, and high feeding rates), or we can do this with a single command, using the tapply() function.

```
> tapply(fish.stacked$Mass, fish.stacked$Trmt, shapiro.test)
```

If you run this, you'll see that it performs the Shapiro-Wilk test on each of the three treatment levels separately. All of the samples are statistically normally distributed, so we may proceed with the parametric ANOVA test. In addition, we should check that the variances of our samples are similar. We can do this using the Bartlett test.

```
> bartlett.test(fish.stacked$Mass, fish.stacked$Trmt)
```

```
        Bartlett test of homogeneity of variances

data:   fish.stacked$Mass and fish.stacked$Trmt
```

```
Bartlett's K-squared = 0.084655, df = 2, p-value =
0.9586
```

The result here suggests that the variances are not significantly different.

There's one more check we should be sure to make. It's important that our data are numeric while our treatment levels are of type "factor." We already know that the data are numeric after testing them for normality and equal variances (**R** would have been more than happy to provide us an error message). We can test the treatment factor to make sure it's stored as a factor, as follows:

```
> class(fish.stacked$Trmt)
```

```
[1] "factor"
```

Sometimes our data will be coded with treatment levels labeled as numbers (e.g., 1, 2, 3, and so on). If this is the case, then **R** interprets those as numeric data instead of different levels of a factor. When this happens, we need to type-cast them into factors using the `as.factor()` function. If they need to be type-cast, then we would use the following code:

```
> fish.stacked$Trmt = as.factor(fish.stacked$Trmt)
```

Let's now write down the hypotheses we are testing using words:

H_0: Samples all come from the same population.
H_A: Samples do not all come from the same population.

Note that if we have seventeen samples and all except one are similar, the ANOVA will lead us to reject the null hypothesis. Let's now complete the analysis. We will store the output from the `aov()` function in a variable for later use.

```
> fish.aov = aov(fish.stacked$Mass ~ fish.stacked$Trmt)
```

We can now let **R** summarize the output for us using the `summary()` function:

```
> summary(fish.aov)
```

```
                  Df  Sum Sq  Mean Sq  F value   Pr(>F)
fish.stacked$Trmt  2   376.6   188.32    11.76  0.000848 ***
Residuals         15   240.2    16.01
---
Signif. codes:
0 '***' 0.001 '**' 0.01 '*' 0.05 '.' 0.1 ' ' 1
```

8.2 Interpreting Results from a One-Way ANOVA

The summary provided above is called an ANOVA table. At first, we usually are most interested in whether the samples are statistically different. We see this, like usual, by inspecting the p-value. This is found on the right side of this table under the Pr (>F) heading, where we see p = 0.0008. Therefore, we can conclude that the masses of fish grown under the different feeding treatments are statistically different ($F = 11.76$; df = 2, 15; $p < 0.001$). Note that we have two degrees of freedom because there are three samples.

Another question we'd like to answer is, Is the effect very strong? Under the column "Sum sq" (sum of squares), we see that the proportion of variance explained by the treatment is $376.6/(376.6 + 240.2) = 0.61$, or 61%, which is most of the variance. You also can get this value, called R squared (R^2), by using the linear model function (the result is not shown):

```
> summary(lm(fish.stacked$Mass ~ fish.stacked$Trmt))
```

Although we've found that the samples are statistically different, we don't know which sample might be different from the others. We need to do a little more work to interpret these more complicated results. For instance, increasing feeding from low to medium rates seems important, but increasing from medium to high rates seems to have a smaller effect, based on the data in figure 8.1. If the ANOVA is statistically significant, we can test for differences among levels in an ANOVA, performing what's called a post hoc test (i.e., after-the-fact test). I often use Tukey's Honest Significant Differences test (the TukeyHSD() function in **R**). It's a good post hoc test if the variances of the different samples are similar, which is an assumption of the ANOVA. Other tests could be used instead. You should have a good reason to choose a different one and always use only one! To implement this test, we send the variable fish.aov to the TukeyHSD() function:

```
> TukeyHSD(fish.aov)
```

```
   Tukey multiple comparisons of means
     95% family-wise confidence level

Fit: aov(formula = fish.stacked$Mass ~ fish.stacked$Trmt)

$'fish.stacked$Trmt'
              diff        lwr       upr      p adj
Med-Low   6.966667  0.9652966 12.96804 0.0223110
High-Low 11.083333  5.0819632 17.08470 0.0006483
High-Med  4.116667 -1.8847034 10.11804 0.2091488
```

The interpretation of the output from this test is a bit challenging and gets more so with increasing numbers of samples. The function provides a table (above) with each pairwise comparison on the left (e.g., Med-Low) with a difference between the means (diff). It also provides lower (lwr) and upper (`upr`) bounds. If those bounds include 0, then the two samples are not statistically different. We also get an associated p-value (`p adj`) for each comparison in the right column. In this example, two of the three comparisons are significantly different ($p \leq 0.05$). However, we see that the medium and high treatment groups are not statistically different.

8.3   Visualizing a One-Way ANOVA

Now we're ready to create a final visualization of our results. We're going to make both a boxplot and a barplot with 95% confidence intervals *and* add contrast letters that show which means are statistically similar or different from each other based on the post hoc Tukey HSD analysis. Your boss/professor/lab instructor may have a preference for the way that they want these results presented, so your job is to be able to do this either way.

We want to identify the groupings of samples using lowercase letters. Samples that are not significantly different from each other share a letter. Because we're going to add letters to these plots, we need to set up a variable that will hold those letters in the correct order (from left to right). To do this, we should build the letters upward in alphabetical order, starting with the lowest mean or median. Therefore, for this example, the correct order of letters is as follows:

```
> contrast.labels = c("a","b","b")
```

This will tell the reader that the masses of fish receiving the low feeding rate are significantly less than the masses of fish receiving the medium and high feeding rates but that the medium and high treatment fish are not significantly different from each other.

Adding Constrast Letters to a Boxplot

Contrast letters should be placed in the middle-top of each sample, which means we need to know where that is for each sample. The function we'll use to add the letters is, not surprisingly, called `text()`. We first make the boxplot graph, find the tops using the `max()` function, and then place the letters *above* the highest value. Note that the command "`pos = 3`" places the letter in position 3, based on the maximum values' locations. Positions are below (= 1), to the left (= 2), above (= 3), and to the right (= 4) (see figure 8.2).

```
> boxplot(fish.stacked$Mass ~ fish.stacked$Trmt, # note the
> tilde
+         names = c("L","M","H"), ylim = c(0,80),
+         xlab = "Feeding Rate", ylab = "Fish Mass (g)",
+         las = 1, cex.lab = 1.5)
```

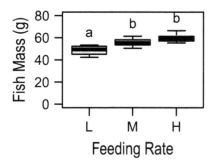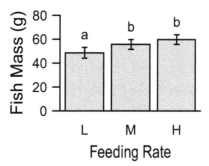

Figure 8.2. The mass of fish grown under low (L), medium (M) and high (H) feeding rates. The data have been graphed using the boxplot() (left) and barplot() (right) functions. We see that fish attain a larger mass when they are fed at a higher rate (F = 11.76; df = 2, 15; p < 0.001). Error bars on the right represent 95% confidence intervals. Samples with different letters are statistically different.

```
> data.tops = tapply(fish.stacked$Mass, fish.stacked$Trmt, max)
> text(1:3, data.tops, labels = contrast.labels, pos = 3)
```

Adding Constrast Letters to a Barplot That Has Error Bars

This is a little harder because we need to calculate and place 95% confidence intervals on each sample mean and then place letters. Also, unlike the logical placement of samples in a boxplot (integer values starting at 1), **R** places the bars at non-integer locations along the x-axis. Therefore, we need to capture the locations for the bars from the barplot() function. We can do this by using a variable, such as "a" below.

The confidence intervals are added using the plotCI() function, which is in the plotrix package (you may need to install that if you haven't before). This function takes as arguments the x-locations ("a," in this example), the height of the bars (the means), the size of the error bars (CI95), and the character to use where the error bar meets the top of the bar (I like having no character using pch = NA), and finally we need to tell the function to add these error bars to the active graph (add = T) (see the right panel in figure 8.2).

```
> library(plotrix)
> M = tapply(fish.stacked$Mass,fish.stacked$Trmt,mean)
> S = tapply(fish.stacked$Mass,fish.stacked$Trmt,sd)
> SEM = S/sqrt(6) # because S is an array, SEM is an array
> CI95 = qt(1-0.05/2,5) * SEM
> a = barplot(M, ylim = c(0,80), las = 1, cex.lab = 1.5,
+             names.arg = c("L", "M", "H"),
+             xlab = "Feeding Rate", ylab = "Fish Mass (g)")
```

```
> abline(h = 0)
> plotCI(a,M,CI95, add = T, pch = NA)
> text(a, M+CI95, labels = contrast.labels, pos = 3)
```

What If the Data are Really Wonky?

Sometimes your data are normally distributed, which allows us to use parametric tests (e.g., ANOVA). Sometimes the data are not normally distributed, and we can use non-parametric tests. However, even non-parametric tests assume the data have all been drawn from the same distribution. Generally, that assumption is valid. For example, the height of corn plants may not be normally distributed, or you're not able to transform them so they are normally distributed, but they're all corn plants and their growth responses should be similar. In the event your data are all over the place, you may have to consider that something really is wrong with your data or that, if your data are correct, you can't do a standard statistical test as described here. Instead, you may need to turn to alternative methods, such as randomization tests.

8.4   One-Way Test for Non-Parametric Data

If the samples are not normally distributed, or can't be made normally distributed with a log-transformation, then we can conduct a Kruskal-Wallis test (but see box 8.2). Keep in mind that this test does not assume samples are normally distributed, but it does assume that the samples come from the same population (the standard H_0). Therefore, all the samples are assumed to come from the same shaped distribution. If you sense that the distributions of the samples are similar, then you may proceed with this test. Below are data for tail lengths in centimeters for three species of small mammals:

```
> sp.A = c(7.03, 7.22, 7.16, 7.15, 7.33, 7.22, 7.77,
+          7.75, 7.33, 7.17)
> sp.B = c(5.84, 5.59, 5.38, 5.38, 5.35, 5.87, 5.29,
+          5.75, 5.85, 5.33)
> sp.C = c(7.09, 7.49, 7.06, 7.19, 7.10, 7.22, 7.45,
+          6.20, 6.96, 6.02)
```

We should combine these into a dataframe and stack them for graphing and conducting our analysis:

```
> tail.dat = data.frame(sp.A, sp.B, sp.C)
> tail.dat = stack(tail.dat)
> names(tail.dat) = c("length","species")
```

Box 8.2. Normally distributed data and ANOVA. Sometimes we'll have data where some samples are normally distributed and one pesky sample is not. What should you do? First, extensive research on ANOVA suggests that it performs well (gives the right answer) even when samples are not normally distributed (it is robust to violations of normality). So we can use ANOVA if a few of our samples are, let's say, kind of not normally distributed. My suggestion is as follows. You can use ANOVA if some of the samples are not normally distributed but:

1. They have somewhat similar distributions to the samples that are statistically normally distributed, or
2. Past experience or information has demonstrated that the process that generated the data should result in data that are normally distributed.

If the data are not normally distributed but you have reason to believe the data should be normally distributed you should seek the advice of a statistician.

Now let's create an appropriate graph to view the distributions of our three samples and compare them side-by-side (see figure 8.3). Note that for a final visualization, we want to add contrast letters if the samples are statistically different from each other.

```
> boxplot(tail.dat$length ~ tail.dat$species,
> names = c("A","B","C"),
```

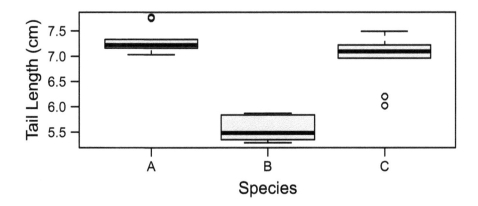

Figure 8.3 Side-by-side boxplots of the distributions of tail lengths for three species. The three samples are statistically different using a Kruskal-Wallis rank sum test ($\chi^2 = 20.6$, df = 2, p < 0.001).

```
+               ylab = "Tail Length (cm)", las = 1,
+               xlab = "Species",cex.lab = 1.5)
```

It's hard to tell from these distributions whether they're normally distributed. Let's test our samples for normality using the Shapiro-Wilk test:

```
> tapply(tail.dat$length, tail.dat$species, shapiro.test)
```

If you run this code, you will find that these samples are not normally distributed. Let's check whether log-transforming these data makes them normally distributed:

```
> tapply(log(tail.dat$length), tail.dat$species, shapiro.test)
```

Again, if you run this, you will find the logged data also are not normally distributed. Therefore, we will proceed with the non-parametric Kruskal-Wallis test for comparing these samples. This test is run with the following function using the formula approach:

```
> kruskal.test(tail.dat$length ~ tail.dat$species)
```

```
        Kruskal-Wallis rank sum test

data:   tail.dat$length by tail.dat$species
Kruskal-Wallis chi-squared = 20.631, df = 2, p-value
= 3.311e-05
```

From this analysis, we find that our data do not support the null hypothesis that the samples come from the same population. Instead, we find that tail lengths for the three species measured are statistically different (χ^2 = 20.6, df = 2, p < 0.001).

If you need to conduct a post hoc test for the Kruskal-Wallis test, you can do this using the `kruskalmc()` function found in the `pgirmess` package. Note that it would be wildly inappropriate to graph these data with a barplot showing means and 95% confidence intervals because both measures assume that the data are normally distributed.

8.5   Two-Way ANOVA

In this section, we discuss a more complicated design and introduce only the test that assumes the data are normally distributed. What do we mean by a two-way (or two-factor) test? Sometimes we are interested in testing whether two (or more!) factors (perhaps watering rates and nutrients) interactively affect our system (such as plant growth). This seems really complicated. You might be tempted to approach this two-factor problem as two separate experiments. For instance, we could test whether different water levels affect plant growth. We might find that plants grow more with more water. Likewise, we might want to test the effect of fertilizer levels on plant growth. Again,

Table 8.1 Design for 2 × 2 factorial experiment to determine the effects and interaction of water and nutrient levels.

Number of Plants	Water Level	Nutrient Level
8	Low	Low
8	Low	High
8	High	Low
8	High	High

we might find that plants grow more with more fertilizer. Why consider a single, two-factor test? Here are two important reasons:

1. We want to know if our factors *interact*.
2. Experiments are expensive and we want to save money. One experiment is more efficient than two separate experiements.

The most important reason for doing a single experiment is to test the interactive effects of multiple factors. It turns out that plants often will respond to fertilizer only if they have a lot of water. If they have low levels of water available, then fertilizing them can actually kill them! So the growth rate of our plants can depend not only on water and fertilizer levels but also on the *interaction* between water and fertilizer. Just to restate this problem: if we test two factors in separate experiments, we would be unable to discover that the two factors might interact. A multifactor design (also called a factorial design) is more complicated but allows us to ask more interesting questions in biology. If we shy away from this complexity and avoid complicated statistical tests, then we may easily miss the really exciting aspects of biological systems.

Designs can actually be much more complicated than just one or two factors, and this complexity is where we are in biology! Unfortunately, the simple questions have been asked and answered and, therefore, are no longer fundable or publishable. So buckle up!

The design shown in table 8.1 is a 2 × 2 factorial design and, as shown, requires 32 plants (2 × 2 × 8 replicate plants per treatment). Here are the data for dried plant masses in grams.

```
> LW.LN = c(3.84,4.47,4.45,4.17,5.41,3.82,3.83,4.67)
> LW.HN = c(8.84,6.54,7.60,7.22,7.63,9.24,7.89,8.17)
> HW.LN = c(7.57,8.67,9.13,10.02,8.74,8.70,10.62,8.23)
> HW.HN = c(16.42,14.45,15.48,15.72,17.01,15.53,16.30,15.58)
```

The variables are each named with four letters (e.g., LW.LN). The first two letters represent the water levels (low water (LW) or high water (HW)), and the last two letters represent the nutrient levels (either low nutrient (LN) or high nutrient (HN)). Now we will build the dataframe for these data using the function rep(), which stands for "repeat."

```
> water = rep(c("LW","HW"), each = 16)
> nutr = rep(c("LN","HN"), each = 8, times = 2)
> plant.mass = c(LW.LN,LW.HN,HW.LN,HW.HN) # combine the data
```

We can see what water and nutr now look like by printing them to the console:

```
> water
```

```
 [1] "LW" "LW" "LW" "LW" "LW" "LW" "LW" "LW" "LW" "LW" "LW"
[12] "LW" "LW" "LW" "LW" "LW" "HW" "HW" "HW" "HW" "HW" "HW"
[23] "HW" "HW" "HW" "HW" "HW" "HW" "HW" "HW" "HW" "HW"
```

```
> nutr
```

```
 [1] "LN" "LN" "LN" "LN" "LN" "LN" "LN" "LN" "HN" "HN" "HN"
[12] "HN" "HN" "HN" "HN" "HN" "LN" "LN" "LN" "LN" "LN" "LN"
[23] "LN" "LN" "HN" "HN" "HN" "HN" "HN" "HN" "HN" "HN"
```

The variable plant.mass contains the combined data as an array. Now we can create the dataframe, using the function data.frame(), and store the data in a variable called plant.dat. We can send this function our three arrays of data from above, and it will assemble everything into one structure.

```
> plant.dat = data.frame(water,nutr,plant.mass)
```

We can view the beginning of this dataframe with the head() function, as follows:

```
> head(plant.dat) # view the first 6 rows
```

```
  water nutr plant.mass
1    LW   LN       3.84
2    LW   LN       4.47
3    LW   LN       4.45
4    LW   LN       4.17
5    LW   LN       5.41
6    LW   LN       3.82
```

We see that the two factors (water and nutr) are at the top of the columns. Our data are in the plant.mass column. We should look at the whole dataframe to make sure each value is correctly placed into its treatment level. We can now create a boxplot of the data using the formula approach (see figure 8.4). Note that we have yet to conduct a hypothesis test.

```
> boxplot(plant.dat$plant.mass ~ plant.dat$water *
> plant.dat$nutr,
+         ylab = "Biomass (g)", xlab = "Treatment",
+         las = 1, cex.lab = 1.5)
```

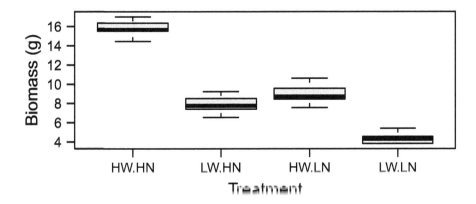

Figure 8.4 A quick visualization using boxplots of our samples. We see the samples all appear relatively normally distributed. Note the naming scheme, which uses the first two letters to represent the water treatment level and the latter two letters to represent the nutrient levels.

Now that we've seen what our results look like, it's time to begin our analysis. H_0 is that the samples have all been drawn from a single population (that water and nutrient levels have no effect on plant mass). That doesn't look like the case in figure 8.4, but we need some quantitative support for our rejection of H_0.

The ANOVA is a parametric test and assumes that the data are normally distributed. Despite the ANOVA's apparent robustness (see box 8.2) we still should test our samples separately for normality. Pulling out just our samples from a dataframe requires subsetting our data. Fortunately, we've done subsetting before (see section 3.5). Here's another way we can subset our data. Let's imagine that we want just the plant masses for the treatment when water was low and nutrient levels were low. This can be done using indices in the dataframe, as follows:

```
> plant.dat$plant.mass[plant.dat$water == "LW" &
+                      plant.dat$nutr == "LN"]
```

```
[1]  3.84 4.47 4.45 4.17 5.41 3.82 3.83 4.67
```

What we're asking **R** to do is to print to the screen the plant.mass data from the dataframe plant.dat. We don't want everything. Instead, we want only the plant.mass when the water level is equivalent to LW (that's the == combination), which is low water *and* (that's the & symbol) when the nutrients are low (LN).

Now that we know how to subset our dataframe, we can send those data as samples to the shapiro.test() function to test for normality.

I have asked **R** to provide just the p-values for these tests by adding the "$p.value" to the function call.

```
> shapiro.test(plant.dat$plant.mass[water == "LW" &
+                         nutr == "LN"])$p.value
```

```
[1] 0.1734972
```

```
> shapiro.test(plant.dat$plant.mass[water == "LW" &
+                         nutr == "HN"])$p.value
```

```
[1] 0.9509188
```

```
> shapiro.test(plant.dat$plant.mass[water == "HW" &
+                         nutr == "LN"])$p.value
```

```
[1] 0.6769018
```

```
> shapiro.test(plant.dat$plant.mass[water == "HW" &
+                         nutr == "HN"])$p.value
```

```
[1] 0.744246
```

As you might notice, pulling out the individual samples is a little tough. I could have done this more simply with just testing each individual variable that we had before joining them into a dataframe (e.g., LL), but it's important to know what to do if you have stacked data. Because the data are normally distributed, we may proceed with the ANOVA hypothesis test.

I can now run the ANOVA with the following call, using the aov() function and storing my result in a variable called plant.aov:

```
> plant.aov = aov(plant.dat$plant.mass ~
+               plant.dat$water * plant.dat$nutr)
```

The model that we want to fit is sent to the aov() function as $y \sim x_1 * x_2$. The multiplication on the right-hand side tells **R** that we want the main effects and the interaction term.

R doesn't report anything because the output is, instead, stored in the variable plant.aov. We should use this approach of saving the output from the aov() function so we can then get and format the output in more useful ways. You can see the results by typing plant.aov at the command prompt. You can have **R** format the output for you by sending the variable plant.aov to the summary() function. **R** will return the ANOVA table.

```
> summary(plant.aov)
```

	Df	Sum Sq	Mean Sq	F value
plant.dat$water	1	314.88	314.88	488.69
plant.dat$nutr	1	216.74	216.74	336.37
plant.dat$water:plant.dat$nutr	1	21.68	21.68	33.65

Table 8.2 ANOVA table for the effect of water and nutrient levels on plant biomass.

	Df	Sum Sq	Mean Sq	F value	Pr(>F)
plant.dat$water	1	314.88	314.88	488.69	0.0000
plant.dat$nutr	1	216.74	216.74	336.37	0.0000
plant.dat$water:plant.dat$nutr	1	21.68	21.68	33.65	0.0000
Residuals	28	18.04	0.64		

```
Residuals                        28  18.04     0.64
                                  Pr(>F)
plant.dat$water                  < 2e-16 ***
plant.dat$nutr                   < 2e-16 ***
plant.dat$water:plant.dat$nutr 3.13e-06 ***
Residuals
---
Signif. codes:
0 '***' 0.001 '**' 0.01 '*' 0.05 '.' 0.1 ' ' 1
```

The output from **R** is a bit messy and awkwardly wrapped here, but the information is important for us to understand what happened in our experiment. I've reformatted the output into table 8.2. In the next section, we'll discuss in greater detail what this means.

8.6   Interpreting the ANOVA Table

As with our other statistical tests, we are quite interested in the p-value so our eyes naturally go to the far right column in table 8.2. We see the p-value is in the column labeled Pr(>F). Notice that all the p-values are zero! Well, they are *not* zero; they are just small. We should report them as $p < 0.001$ in a laboratory report (you might revisit our discussion of p-values in section 6.8).

There are three effects in the table, one for each of our two factors, called main effects, and a third for the interaction effect between these two factors. The first main effect is for the effect of water on plant mass. This line gives us the information we need to interpret the effect of our watering levels on plant biomass. Go across this row to see that the degrees of freedom (Df) for this term is 1. This tells us the number of levels, which is one more than Df (Df = $n - 1 = 1$). So there must have been two levels ($n = 2$). The mean square error term (Mean Sq) is the sums of squares (Sum Sq) divided by the degrees of freedom (Df). The F value is the Mean Sq term divided by the residuals Mean Sq term in the Residuals row (0.64). The F statistic is used to determine a p-value that, in **R**, is labeled "Pr(>F)."

In general, the bigger the F value, the smaller our p-value. It's our job to compare the p-values against our pre-determined α, which we usually set at 0.05, to determine statistical significance.

The second row is the main effect for the nutrient treatment. The third row represents the interaction term for the water by nutr treatment effect. Some researchers consider the interaction term to be more important than the main effects and even use a different (usually higher) critical p-value (α) to assess the significance of this effect. Check with your laboratory instructor or mentor about how they want you to interpret higher-order effects. Here, we'll be conservative and stick with $\alpha = 0.05$.

Our results suggest strong main effects for both water level and nutrient level on plant biomass. In general, we see that plants with high nutrient levels (the two boxplots on the left in figure 8.4) are larger, on average, than plants with low nutrient levels (the two boxplots on the right in figure 8.4). We also see that plants with higher water levels (HW) tend to be larger than plants with lower water levels (LW). Does that make sense? It's always important, and sometimes tricky, in these tests to know what the results should look like. Be careful to not just plug and chug statistical tests. Work to understand what these results tell us about the data we see in our graph.

The interaction term is usually the really confusing part of this analysis but also the most important part. To help us interpret this, we can use the interaction.plot() function (see figure 8.5 for a description of what

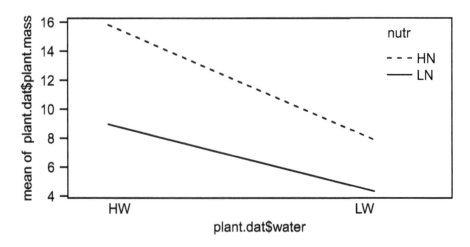

Figure 8.5 An interaction plot for the plant biomass analysis. Notice that, if the lines connecting the means of the nutrient factor (high and low nutrients in this example) are parallel, this suggests there is no statistical interaction. If the lines are not parallel, it suggests there is a significant interaction. In this graph, the lines are not parallel, which agrees with the output from the ANOVA test (F = 33.65; df = 1, 28; p < 0.001).

this plot shows). This is an important graph for visually inspecting statistical significance. If the interaction term is not significant, these lines will be approximately parallel. In this example, the lines are not parallel, which is consistent with the significant water by nutrient interaction effect.

```
> interaction.plot(plant.dat$water, trace.label = "nutr",
> las = 1,
+       trace.factor = plant.dat$nutr, plant.dat$plant.mass)
```

To determine whether individual samples differ from each other, we need to use a post hoc comparison (discussed above in section 8.2). Here's how to do this with our data.

```
> TukeyHSD(plant.aov)
```

```
  Tukey multiple comparisons of means
    95% family-wise confidence level

Fit: aov(formula = plant.dat$plant.mass ~ plant.dat$water *
        plant.dat$nutr)

$'plant.dat$water'
          diff       lwr       upr p adj
LW-HW -6.27375 -6.855088 -5.692412     0

$'plant.dat$nutr'
         diff       lwr       upr p adj
LN-HN -5.205 -5.786338 -4.623662     0

$'plant.dat$water:plant.dat$nutr'
                  diff        lwr        upr       p adj
LW:HN-HW:HN   -7.92000  -9.01582151  -6.824178 0.0000000
HW:LN-HW:HN   -6.85125  -7.94707151  -5.755428 0.0000000
LW:LN-HW:HN  -11.47875 -12.57457151 -10.382928 0.0000000
HW:LN-LW:HN    1.06875  -0.02707151   2.164572 0.0579272
LW:LN-LW:HN   -3.55875  -4.65457151  -2.462928 0.0000000
LW:LN-HW:LN   -4.62750  -5.72332151  -3.531678 0.0000000
```

The output from this test at first appears to be quite confusing. There's a lot of information here. We see the results for the two main effects (water and nutr) individually, followed by the interaction comparison (below the $'plant.dat$water:plant.dat$nutr'). For the interaction comparison, each of the six rows represents a statistical, pairwise test between two treatments. The first row, for instance, is for the LW:HN-HW:HN comparison. This is a comparison between the sample of plants that receive low water and high nutrients (LW:HN) against the sample of plants that receive high water and high nutrients (HW:HN). The last column shows the result of the statistical test ("p adj"), which is the p-value. You can see that they are all statistically significant (p < 0.001) except for one comparison, which has p = 0.0579. Can you find that comparison in figure 8.5?

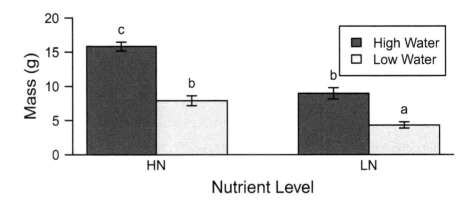

Figure 8.6 A barplot of the means for a two-way ANOVA for plant response to water and nutrient treatments. Error bars are ± 95% confidence intervals. Samples that share a letter are not statistically different.

This post hoc test on the interaction term *can only be assessed if the interaction term is statistically significant.* Once we do this post hoc test, we can interpret which of the samples differ from each other.

If the relationship is significant, we need to determine letters that tell the reader which samples are statistically similar or different from each other. Samples that share a letter are not statistically different. Begin with the sample with the smallest mean and work upward alphabetically. Below are the contrast letters that are consistent with this approach. In the next section, we'll discuss how to add the following contrast letters to figure 8.6.

```
> contrast.labels = c("c","b","b","a")
```

8.7   Visualizing the Results of a Two-Way ANOVA Test

Interpreting a two-way ANOVA, as we have seen, can be tricky. The interaction plot (figure 8.5) helps us to understand the results from our experiment but falls short of a professional-quality graph. Instead, the standard practice is to use a barplot. I've added 95% confidence intervals to this graph to represent the variability in each sample. You may be asked to use a different measure of variability that uses the same approach as the next code section, which uses the `tapply()` function. To build this graph, we need to have all our summary statistics in their own two-dimensional matrix. This way we can group the bars by the factor of our choice and include the matching error bars.

We need to gather the means and confidence intervals for our data and store the results in separate matrices. The `tapply()` function performs a cross-tabulation on the data and applies the appropriate function (e.g., the `mean()`) to the data by groups. We can store the means in a matrix called M to create our barplot, as shown in figure 8.6 (also see the description of making barplots in chapter 5). We gather our estimates of variability similarly to calculate the 95% confidence intervals (see code below). When we make the barplot, we again need to capture the x-axis locations of each bar, so I keep those returned values in a variable called "a." Note that you need to have installed the `plotrix` package in order to use the `plotCI()` function. The last line of code adds the contrast letters as text on the top of the error bars using the `text()` function.

```
> M = tapply(plant.dat$plant.mass,
+            list(plant.dat$water, plant.dat$nutr), mean)
> SD = tapply(plant.dat$plant.mass,
+            list(plant.dat$water, plant.dat$nutr), sd)
> SEM = SD/sqrt(8) # there are 8 observations per sample
> CI95 = qt(0.975,7)*SEM
> a = barplot(M, beside = TRUE, ylim = c(0,20),
+   xlab = "Nutrient Level", ylab = "Mass (g)",
+   legend = c("High Water","Low Water"),
+   cex.lab = 1.5, las = 1)
> abline(h = 0)
> library(plotrix)
> plotCI(a, M, CI95, pch = NA, add = T)
> text(a, M + CI95, labels = contrast.labels, pos = 3)
```

If we want to switch the factor on the x-axis and the trace factor (water, which shows up in the legend), then we just reverse the order of the factors in the list in the `tapply()` function and change the legend text to be high and low nutrients.

Writing the Results for a Two-Way ANOVA

This task is complicated enough to warrant a small section. In the previous example, we have a variety of results. We have the two main effects for water and nutrient addition on plant mass. In addition, we have the interaction term. Finally, the interaction term is significant, so we conducted a post hoc Tukey HSD test and found some differences among those samples. It is important to report the ANOVA table. It should look something like table 8.2. You should clean up the names of the treatments and make sure that the p-values are changed to "< 0.001." Once you've provided the ANOVA table, most of your work is done.

Each of the three effects in the table warrants a sentence of what you found. Start by telling the reader what the main effects mean (e.g., adding nutrients increased plant mass). This should be done using active voice

("I found ..."), unless someone instructs you otherwise. Finally, end the sentence with the pertinent statistical output. For example, "I found that adding water increased plant mass significantly (F = 488.7; df = 1,28; p < 0.001)." A common mistake is to state that the treatment had a significant effect, which doesn't tell the reader what really happened. In addition, if you have a figure for a result, you should refer the reader to it.

Finally, if the interaction term is significant, then you have a challenge. In the plant example, shown in figure 8.6, we should discuss what we find most interesting. It might be the samples that are not statistically different. Or it might be that high water and high nutrient addition acted synergistically to increase plant mass. Whatever you choose to point out, your sentence about it should end with the statistical output for that interaction (i.e., F = 33.6; df = 1, 28; p < 0.001).

8.8   Problems

1. Table below shows data for the bacterial counts found in milk from different dairies. The columns represent different farms, and the rows are separate replicates from each farm (six from each of the five farms). Units are in thousands of colony-forming units per milliliter (cfu ml^{-1}).

   (a) Enter the data below into an Excel spreadsheet as they appear. For instance, in Excel, enter "Farm 1" in cell A1, the number 24 in A2, and so forth. Save this file in the .csv format.
   (b) In your script file in RStudio, read the data file into a dataframe variable called milk.dat.
   (c) Stack the data into the variable milk.dat.stacked. Check that these are correct by sending the output from the variable to the console.

Farm 1	Farm 2	Farm 3	Farm 4	Farm 5
24	14	11	7	19
15	7	9	7	24
21	12	7	4	19
27	17	13	7	15
33	14	12	12	10
23	16	18	18	20

(d) Test whether the data are normally distributed. Remember that you need to test the counts within each farm separately. Show your code and the results.

(e) Test whether there is a statistical difference in colony-forming units among the farms.

(f) Create a barplot with 95% confidence intervals of the bacterial counts as a function of the farm. If there is a statistical difference among farms, add appropriate letters to each sample indicating those differences.

(g) Provide a clear results statement for your finding and include, parenthetically at the end of your statement, the statistical support for your conclusion.

2. Below are data on the lengths of twenty stickleback fish grown under four different conditions (cold/warm water and high/low pH tanks).

(a) Enter these data into a spreadsheet with three columns. These columns should be labeled as "Temp," "pH," and "Length." Save them in the .csv format. Read the spreadsheet into an **R** dataframe called my.sticks. Verify that they are entered correctly in **R** by typing my.sticks at the console and check them against the data in the table.

(b) Test whether the samples are normally distributed.

(c) Assume that the data are normally distributed and provide an ANOVA table of the main and interaction effects.

(d) Provide two, separate side-by-side barplots of the means for the main effects (temperature and pH). Include 95% confidence interval error bars for both graphs. Note that you don't need to add letters when there are just two samples; they either are or are not significantly different.

Temp	pH	Length (cm)	Temp	pH	Length (cm)
C	H	4.2	W	H	4.4
C	H	5.0	W	H	4.3
C	H	4.4	W	H	4.4
C	H	4.4	W	H	4.5
C	H	4.1	W	H	4.6
C	L	3.2	W	L	5.6
C	L	3.2	W	L	6.1
C	L	3.1	W	L	5.9
C	L	2.3	W	L	6.9
C	L	3.4	W	L	5.7

(e) Provide a single barplot of means for the interaction term for this analysis. Include **95%** confidence intervals along with letters indicating significant differences, determined through the use of the Tukey Honest Significant Difference post hoc test.

(f) Provide a paragraph summarizing the results from this analysis. Note that you have two main effects and an interaction term to consider. Be sure to end each statement with the appropriate statistical support.

CHAPTER NINE

Hypothesis Tests: Linear Relationships

WE ARE OFTEN INTERESTED in asking whether two variables are related. For instance, we might ask whether the mass of dogs is related to the length of dogs. In this kind of test, the data are generally continuous and paired together somehow (a dog has a mass and a length). Two measurements from the same experimental unit? It's okay because we use each dog (length and mass) as a single datum (point). We do need to have independence, however, between our data points. We wouldn't want to test this relationship between dog lengths and masses using only dogs from a single litter (they wouldn't be independent).

There are two broad types of tests we might perform on such data, assuming we're investigating whether there is a linear relationship between two variables. These tests are correlation and regression analyses. We perform a correlation analysis when we're interested in determining if there is a linear relationship between two normally distributed variables, and a regression analysis when we're interested in testing whether one continuous variable is dependent on the other variable. Strangely, the correlation test assumes a linear relationship, but we never add a best-fit line to the data! This can cause a lot of confusion about whether you should add a line to a scatterplot. After working through this chapter, you will know when adding a line is appropriate.

9.1   Correlation

As discussed above, a correlation analysis is a test that investigates whether two variables are linearly related. Let's assume we are interested in the relationship between the number of bars and the number of churches in a variety of towns across New York State. Here are my data for ten towns of different sizes:

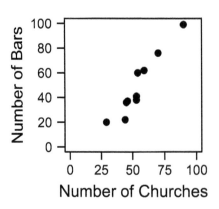 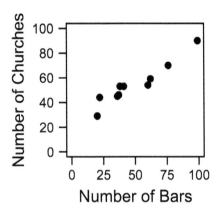

Figure 9.1 The relationship between number of churches and the number of bars in ten towns of different sizes. There appears to be a positive, linear relationship. Does it matter which variable goes on the y-axis?

```
> churches = c(29, 44, 45, 46, 53, 53, 54, 59, 70, 90)
> bars =     c(20, 22, 36, 37, 38, 41, 60, 62, 76, 99)
```

Box 9.1. *Should I add a best-fit line to a scatterplot?*

Correlation. No. There is NEVER a best-fit line added to a correlation plot. There is no causality (or dependence) implied. Either variable can be plotted on the y-axis. A correlation analysis results in a correlation coefficient (r) and a p-value.

Regression. Maybe. The dependent variable (response variable) is placed only on the y-axis. The independent variable (predictor variable) goes on the x-axis. You may add a line to regression data only if the relationship is statistically significant. Provide the reader with the statistical output (F; df_1, df_2; p-value). You also might report the adjusted R^2 value (check with an instructor about what is wanted). If the relationship is statistically significant, add the line to the graph and report the equation (e.g., $y = 2.7x + 14$).

I have graphed them in figure 9.1. There seems to be a positive, linear relationship. We might wonder if we should put a best-fit line through data points (see Box 9.1). What we need to ask is whether it seems the variable on the y-axis is *dependent* on the variable on the x-axis across different towns.

The proper way to approach these data is first to graph them (figure 9.1). I've graphed the scatterplots with the axes presented both ways. Which one is correct? Actually, it would be hard to argue that one variable *depends* on the other. The truth, however, is that both these variables depend on the number of people living in these different municipalities (a third, confounding factor). A confounding factor can also be thought of as a hidden factor.

Assuming we are convinced that neither variable depends on the other and that they appear linear, we can perform the correlation analysis. The correlation analysis is a parametric test and assumes both variables are normally distributed. Below I test this but report only the p-values.

```
> shapiro.test(churches)$p.value
```

```
[1] 0.3819124
```

```
> shapiro.test(bars)$p.value
```

```
[1] 0.3385011
```

Now we may proceed with the correlation test using the `cor.test()` function. Note that the H_0 is that there is no correlation.

```
> cor.test(bars,churches) # Are these correlated?
```

```
        Pearson's product-moment correlation

data:  bars and churches
t = 8.876, df = 8, p-value = 2.052e-05
alternative hypothesis: true correlation is not equal to 0
95 percent confidence interval:
 0.8077248 0.9890719
sample estimates:
      cor
0.9527938
```

R provides us a lot of information from this analysis. We see the type of test (Pearson's product moment correlation test), the test results, and the correlation coefficient (0.953). With this analysis, we conclude that there is a highly significant, positive correlation between the number of bars and the number of churches in ten towns in New York State ($r = 0.953$, $t = 8.88$, $df = 8$, $p < 0.001$).

Correlation with many variables

You can perform a correlation analysis on many variables at one time, but it's important that we focus our hypotheses on what particularly interests us. We must avoid fishing expeditions where we go looking for anything that might be significant. Those efforts are generally considered bad scientific practice and can easily lead to making Type I errors. It is helpful, however, to use the exploratory graphing function `pairs()` to visualize all of the relationships

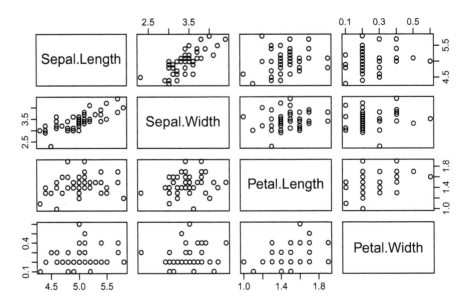

Figure 9.2 A pairs () graph showing all possible scatterplots of the four flower measurements of *Iris setosa*. In the upper-left graph, we see "Sepal.length." This measurement is on the x-axis for the graphs in the left column and on the y-axis for the graphs in the first row.

between all our variables. We can use this approach to seek out relationships that might be of potential interest.

We will look at a built-in dataset that contains information on the flowers of *Iris setosa*. We can get these data using a subsetting technique. The data are in the dataframe "iris" and the species are in the column labeled "species."

```
> flr.dat = subset(iris,Species == "setosa")
```

We can now use the pairs () function to look at the relationships between sepal length and width and petal length and width. Note that our graphing the data does not constitute conducting a hypothesis test. This type of figure is great for data exploration, allowing us to visualize easily whether different variables might be positively or negatively related, or not related at all (see figure 9.2).

```
> pairs(flr.dat[1:4]) # we only need the data
```

We can evaluate these relationships quantitatively by examining the correlation matrix of these data using the cor () function with the following code:

```
> cor(flr.dat[1:4])
```

	Sepal.Length	Sepal.Width	Petal.Length
Sepal.Length	1.0000000	0.7425467	0.2671758
Sepal.Width	0.7425467	1.0000000	0.1777000
Petal.Length	0.2671758	0.1777000	1.0000000
Petal.Width	0.2780984	0.2327520	0.3316300
	Petal.Width		
Sepal.Length	0.2780984		
Sepal.Width	0.2327520		
Petal.Length	0.3316300		
Petal.Width	1.0000000		

The command cor(flr.dat) provides all the pairwise correlation coefficients (r). Notice that on the main diagonal $r = 1.0$. This is because all the variables are perfectly correlated with themselves. At this point, we have *not* conducted a hypothesis test. It's always good to keep the number of statistical tests to a minimum to avoid making statistical errors (see section 6.9). If, after careful consideration, you want to conduct a statistical test and get the p-value for a particular pairing, then run the test. Below I have tested the correlation between sepal length and sepal width:

```
> cor.test(flr.dat$Sepal.Length, flr.dat$Sepal.Width)
```

```
        Pearson's product-moment correlation

data:  flr.dat$Sepal.Length and flr.dat$Sepal.Width
t = 7.6807, df = 48, p-value = 6.71e-10
alternative hypothesis: true correlation is not equal to 0
95 percent confidence interval:
 0.5851391 0.8460314
sample estimates:
      cor
0.7425467
```

For this analysis, we see four important pieces of information. We get a t statistic; df; p; and "cor," the correlation coefficient. From this, we see that the p-value is very small and conclude that there is a positive correlation between sepal length and width in *Iris setosa* ($r = 0.743$, $t = 7.68$, df $= 48$, $p < 0.001$). We note that the degrees of freedom (df) for correlation tests is $n - 2$, that is, the number of observations minus one degree of freedom for each variable.

9.2   Linear Regression

The linear regression is used when you know that y depends on x in a linear fashion (or that relationship has been made linear through a transformation, like logging the data). It assumes the y-variable is normally distributed (see section 4.4).

Some people think the goal of a regression is just to add a line on a scatterplot because the data points look naked. However, the meaning of a best-fit relationship is much more interesting. We often have worked hard and spent time and money to get each value in our dataset. We consider each value as giving us insider information about how the world works. For instance, we might be asking a question about whether something increases, or just changes, as a function of time. This would suggest we're actually interested in knowing the slope of the relationship, which often represents a rate. We also are often interested in where the line crosses the y-axis, called the y-intercept. Our methods here will allow us to formalize both our estimates of these parameters (slope and intercept) and also provide us the estimates of error for these parameters.

Let's consider the example of the relationship between blood alcohol concentration (BAC) and the number of drinks consumed for a 140 lb female. Here are the estimated data (from https://www.tabc.state.tx.us/publications/brochures/BACCharts.pdf):

```
> drinks = 1:10
> BAC = c(0.03, 0.07, 0.11, 0.13, 0.16, 0.19, 0.23, 0.26,
+          0.29, 0.32)
```

The first thing we should consider is that there might be a relationship between these variables. Does having more drinks influence BAC? Probably! If we graph these, which of the variables goes on the y-axis? If BAC depends on the number of drinks someone has, then BAC should go on the y-axis. But can we switch the axes? Can the number of drinks just as easily go on the y-axis? Let's graph both; see figure 9.3.

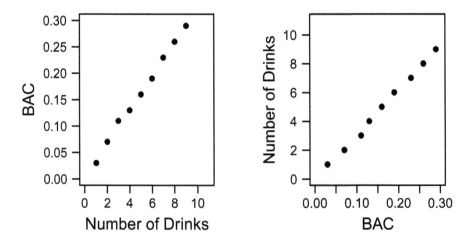

Figure 9.3 The left panel shows blood alcohol concentration (BAC) graphed as a function of the number of drinks for 140 pound females taken in one hour. The right panel switches these axes. Are these both plausible ways to graph these data? Data from https://www.tabc.state.tx.us/publications/brochures/BACCharts.pdf.

```
> par(mfrow = c(1,2))
> plot(drinks,BAC, pch= 16, cex.lab = 1.5, xlim = c(0, 11),
+       ylim = c(0,0.3), xlab = "Number of Drinks", las = 1)
> plot(BAC,drinks, pch= 16, cex.lab = 1.5, ylim = c(0, 11),
+       xlim = c(0,0.3), ylab = "Number of Drinks", las = 1)
> par(mfrow = c(1,1))
```

The only plausible graph is the left panel. BAC depends on how many drinks have been consumed, not the other way around. So we now have a relationship that exhibits a dependency. And the data appear to be linearly related, so we can begin assessing this relationship using linear regression.

Linear regression analysis is a parametric test that assumes that the y axis variable is normally distributed for given values on the x-axis. This is a bit difficult to assess, but we can do this by looking at how the residuals change against the x variable (see box 9.2). We can check that the y-axis data are normally distributed:

```
> shapiro.test(BAC)$p.value
```

```
[1] 0.9199544
```

The p-value is greater than 0.05, which suggests we not reject the H_0 that these data are normally distributed. Therefore, we have more support for conducting a linear regression analysis.

Linear regression will inform us about whether the relationship is statistically significant and whether we can add a line to our graph. For linear regression, we are asking whether the slope of the best-fit line to these data is significantly different from zero ($H_0 : slope = 0$; H_A: slope $\neq 0$). To accomplish this, we use the linear model function lm().

```
> mod = lm(BAC ~ drinks) # this is y ~ x
```

With this type of statistical test, it's good practice to store the result in a variable, such as "mod," so that we can parse out the different pieces. We also can send that result to other functions to do things like get the p-value or draw the best-fit line on our graph. We can get most of what we need using the summary(), so let's look at our results:

```
> summary(mod)
```

```
Call:
lm(formula = BAC ~ drinks)

Residuals:
      Min        1Q    Median        3Q       Max
-0.006364 -0.002773 -0.000697  0.001894  0.010242

Coefficients:
             Estimate Std. Error t value Pr(>|t|)
(Intercept) 0.0046667  0.0034363   1.358    0.212
drinks      0.0316970  0.0005538  57.235 9.64e-12 ***
---
```

```
Signif. codes:
0 '***' 0.001 '**' 0.01 '*' 0.05 '.' 0.1 ' ' 1

Residual standard error: 0.00503 on 8 degrees of freedom
Multiple R-squared:  0.9976,    Adjusted R-squared:  0.9973
F-statistic:   3276 on 1 and 8 DF,  p-value: 9.641e-12
```

Box 9.2. Testing the assumption of linear regression. Warning: this is going to sound crazy! Before we do statistical tests, we need to test the assumptions of the test. However, to test the assumption of linear regression, we're going to conduct the linear regression analysis and assess part of the analysis but *not* look at whether the relationship is statistically significant.

To accomplish this, we run the linear regression analysis on our data and plot the residuals.

```
> mod = lm(BAC ~ drinks)
> plot(mod)
```

In the console, it should say "Hit <Return> to see next plot." Click in the console and hit the "Enter" once. The first graph produced shows the "Residuals" versus the "Fitted Values." This graph should show the following two results that test whether the data are correctly fit by a linear model:

1. The average line through the data should be approximately horizontal along the y = 0 line.
2. The points should be evenly spread above and below this line throughout the range of the fitted values (x-axis).

For this dataset, it's not really that clear because we have only ten data points. It turns out that this passes our test. Below, however, are two examples of residuals graphs that would lead us to reject quickly the assumption for linear regression:

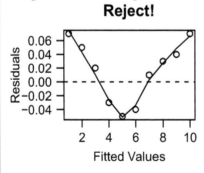

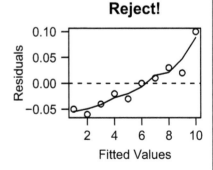

From our analysis, we're interested in getting the following:

1. The equation (or model) that describes the relationship ($y = f(x)$).
2. The significance of the statistical test.
3. How well the points fit the model (the R^2 value).

From the summary above, we can get all of this information. The output, when formatted like this, provides us a section under the "Coefficients," which displays the estimates for the slope, labeled "drinks," and the intercept. We also get the standard errors for these estimated values. The last column gives us the p-value for a hypothesis test for the intercept and slope. H_0 for both of these statistics is that they are zero. We see that the estimate for the intercept is very close to 0.0 and that the resulting p-value is very close to 1.0 (to several decimal places). This strongly suggests that the intercept is not statistically different from zero. That makes sense! If a female (remember the data were for females) does not have a drink, her BAC should be zero.

In addition, the output suggests that the slope is different from zero ($p = 2.318 * 10^{-12}$). This also is the `p-value` for the regression test. When we report the results from a regression, we generally need the F statistic, the two degrees of freedom, the p-value, the adjusted R^2 value, and the equation, and all are provided in this summary.

The larger the F statistic, the more confidence we have that there is a nonzero slope of the y-variable on the x-variable. The R^2 value, also known as the coefficient of determination (also written as r^2), is a value that can range as follows: $0 \leq R^2 \leq 1$. It gives us the proportion of the variance in our y variable that is explained by our x variable. In this particular case, our R^2 value is quite high (almost 1.0), which suggests a very strong fit.

If the slope (or overall regression test) is significant, then we can add the best-fit regression line. The easiest approach is to use the `abline()` function, which draws a line that goes all the way from the axis on the left to the axis on the right (the two ordinates) (see the left panel in figure 9.4). Unfortunately, we shouldn't do this because we need to be very careful about predicting y-values beyond the range of our x-values.

To avoid this we can, instead, use the following code to draw our line, assuming we have our data in two arrays of equal length called x and y:

```
> lines(x,fitted(lm(y~x)))
```

Undoubtedly, this single line of code seems relatively complicated to draw a best-fit line on some data. It has a function (`lm()`) that's in a function (`fitted()`) that's in yet another function (`lines`). After our initial shock that this is what it takes to draw a simple line, let's tear it apart, working from inside out.

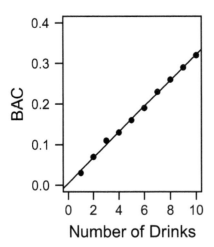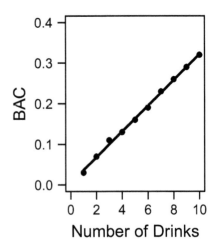

Figure 9.4 The data for blood alcohol content (BAC) graphed as a function of the number of drinks consumed by male subjects. In the left panel, the line *incorrectly* extends beyond the range of the x-variable. In the right panel, the line is graphed correctly, extending only as far as the range of drinks consumed. I've asked the function to double the thickness of the line with the argument lwd = 2.

The fitted() function takes the output from the lm() function and determines the predicted y values for the provided x values. Finally, the lines() function draws the line between all the points (it's straight because all the fitted points lie on the same function returned by lm()). The end product is the graph on the right of figure 9.4.

```
> par(mfrow = c(1,2))
> plot(drinks, BAC, pch= 16, cex.lab = 1.5,
+       xlim = c(0,10), ylim = c(0,0.4),
+       xlab = "Number of Drinks", las = 1)
> abline(mod)
> plot(drinks, BAC, pch= 16, cex.lab = 1.5,
+       xlim = c(0,10), ylim = c(0,0.4),
+       xlab = "Number of Drinks", las = 1)
> lines(drinks, fitted(mod), lwd = 2)
> par(mfrow = c(1,1))
```

One last option to mention is that, with the above example, it is reasonable to assume that the relationship must go through the origin. A person who has not had a drink must have BAC equal to zero. We can force **R** to make the best-fit relationship pass through the origin with one little trick:

```
> mod.origin = lm(BAC ~ drinks - 1)
# -1 force fit through origin
> summary(mod.origin)
```

We should consider this our best model for these data and report our result. BAC for males increases linearly with an increasing number of drinks consumed in one hour (BAC = 0.032 * num.drinks; F = 3276; df = 1, 8; p < 0.001; adj. R^2 = 0.997). Note that I dropped the intercept and did not report it in the equation.

9.3   Another Example of Regression Analysis

A paper from the journal *Science* reported how warning calls of black-capped chickadees (*Poecile atricapillus*) are influenced by the relative perceived threat of a predator (Templeton, Greene, and Davis, 2005). Chickadees are small song birds that make a call that sounds, not surprisingly, like "chick-a-dee." The researchers found that chickadees tag on extra "dee" sounds to their calls depending on the extent of a threat–the more threatened they are, the more "dee" sounds they make. Larger predators actually are less of a threat to chickadees than smaller predators, so the average number of "dee" sounds *decreases* with the size of potential predators. Given this, we should expect to see a graph of the number of calls going down as the size of the predator goes up.

I got the data from figure 9.5 using a website called WebPlotDigitizer (https://apps.automeris.io/wpd/). Here are the data that I was able to get:

```
> aveNumD = c(3.95, 4.08, 1.74, 2.75, 3.03, 3.55, 2.27, 3.16,
+             2.19, 3.21, 2.79, 2.450, 1.34, 2.24, 2.55, 2.05)
> PredLength = c(15.2, 17.5, 22.1, 25.0, 28.5, 31.5, 34.1, 44.2,
+                45.0, 45.0, 47.0, 48.1, 48.5, 52.1, 52.9, 58.0)
```

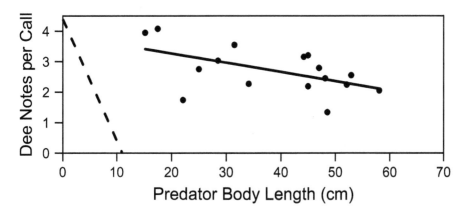

Figure 9.5 The number of "dee" call notes from chickadees in the presence of different potential predators, represented by their average lengths. The solid line is the best-fit line through these data ($y = -0.03x + 3.88$; $F = 5.837$; $df = 1, 14$; $p = 0.03$, adj. $R^2 = 0.244$). The dashed line is the equation presented in the paper ($y = -0.4x + 4.4$, $p < 0.0001$, $R^2 = 0.361$), which is clearly wrong. Data from Templeton, Greene, and Davis (2005).

Before completing our analysis, we must get the linear model and plot the residuals *before* we complete the hypothesis test.

```
> mod = lm(aveNumD ~ PredLength)
> plot(mod)
```

From the residuals graph, which is not shown, the residuals do appear to be relatively flat and evenly spread over the fitted values. Therefore, we can proceed with our hypothesis test, which we complete by sending the model (mod) to the summary() function.

```
> summary(mod)
```

```
Call:
lm(formula = aveNumD ~ PredLength)

Residuals:
    Min      1Q  Median      3Q     Max
-1.4670 -0.3311  0.0270  0.5562  0.7327

Coefficients:
            Estimate Std. Error t value Pr(>|t|)
(Intercept)  3.88097    0.51246   7.573 2.58e-06 ***
PredLength  -0.03050    0.01262  -2.416   0.0299 *
---
Signif. codes:
0 '***' 0.001 '**' 0.01 '*' 0.05 '.' 0.1 ' ' 1

Residual standard error: 0.6628 on 14 degrees of freedom
Multiple R-squared:  0.2942,     Adjusted R-squared:  0.2438
F-statistic: 5.837 on 1 and 14 DF,  p-value: 0.02994
```

```
> plot(PredLength,aveNumD,pch = 16,xlim = c(0,70),
+       ylim = c(0,4.5), las = 1,
+       xlab = "Predator Body Length (cm)",
+       ylab = "Dee Notes Per Call",
+       xaxs = "i", yaxs = "i",cex.lab = 1.5)
> mod = lm(aveNumD ~ PredLength)
> lines(PredLength,fitted(mod),lwd = 2)
> # Add the line from description of Figure2B.
> abline(a = 4.4, b = -0.4, lwd = 2,
+        lty = 2) # line from pub
```

We see that the relationship is statistically significant ($F = 5.837$; $df = 1, 14$; $p = 0.03$, adj. $R^2 = 0.244$), so we can add the best-fit line to our data (see figure 9.5).

As described in the figure caption, there are two lines fit on the graph. The solid line represents the best fit for the data I got using **R**. The dashed line is the equation for the relationship presented by the authors of the paper. It's important to think skeptically. I didn't know the equation was wrong until I plotted it with the data (see figure 2B in Templeton, Greene, and Davis, (2005)).

9.4 Problems

1. Identify whether these examples of data relationships would best be investigated using correlation or regression analysis. Also, identify which variable belongs on the x-axis or if it doesn't matter. Finally, do you expect the relationship to be positive or negative? For such problems, it's usually quite helpful to hand-draw a graph of the expected relationship.

Question	Relationship	Corr/Reg?	x-Axis Data	±
a	The length of metacarpals and metatarsals in humans			
b	Animal speed and mass			
c	Person's life expectancy and cigarettes smoked day^{-1}			
d	Mammalian heart rate and body mass			
e	Number of bird species in 100 hectare plots and plot latitude, north of the equator			

2. The following data are femur and humerus lengths (mm) for five fossils of the extinct proto-bird *Archaeopterix* (Houck et al., 1990). If they are from the same species, then these points should exhibit a highly significant, positive linear relationship. Provide a results statement about whether these data appear to all come from the same species.

Individual	Femur (mm)	Humerus (mm)
1	38	41
2	56	63
3	59	70
4	64	72
5	74	84

   (a) Test the assumptions of correlation or regression (whichever is appropriate) for these data.
   (b) Test whether there is a significant relationship between these variables.
   (c) Graph these data and, if you decide that a line is appropriate, add the best-fit line. If not, do not add the line.
   (d) Explain why a line should or should not be added to the data.
   (e) What does this relationship suggest about these fossils?

3. The following are the number of breeding pairs of red-tailed hawks in a state from 2015 to 2020. As you work with these data, it's interesting to consider whether it is possible that year *causes* the number of pairs to change over time. When we have time series data like these, we recognize that we also can use regression (if appropriate) to determine the rate at

which something is changing. Note that a rate has units of something (e.g., number of breeding pairs) over time. Therefore, time can be an independent factor in a regression analysis.

Year	Number of Breeding Pairs
2015	955
2016	995
2017	1029
2018	1072
2019	1102
2020	1130

(a) Test the assumptions of correlation or regression (whichever is appropriate) for these data.

(b) Test whether there is a significant relationship between these variables.

(c) Graph these data and, if you decide that a line is appropriate, add the best-fit line. If not, do not add the line.

(d) Explain why a line should or should not be added to the data.

(e) If appropriate, what is the rate of change in the number of breeding pairs for this species in this state?

4. Loblolly pine (*Pinus taeda*) trees are found in the southeastern United States. In the built-in dataframe `Loblolly`, you'll find data for the heights of trees at different ages for trees grown from different seed sources.

(a) Subset the dataframe for two "`Seed`" sources. Place the results in two separate, well-named dataframes.

(b) On a single graph, make a scatter plot of the relationship between height and age for both seed types you have chosen. Which of these makes more sense on the x-axis? Explain your reasoning.

(c) Correctly test the relationship for these two seed types. Do they appear to be different? Provide a reason to support your answer.

(d) If appropriate, add lines to your graph for each seed source. If not appropriate, state why you think lines should not or cannot be added.

CHAPTER TEN

Hypothesis Tests: Observed and Expected Values

SOMETIMES WE HAVE DATA where we have counted the numbers of things. Such data are generally called categorical data. A classic example we often run into in biology laboratories is to test whether phenotypes adhere to a 3 : 1 ratio or a 9 : 3 : 3 : 1 ratio. We'll discuss the chi-square test (χ^2 test) and, when the dataset has few observations, the Fisher exact test.

10.1   The χ^2 Test

Imagine you have fifty yellow pea plants and twenty green pea plants. You're wondering if they adhere to a 3 : 1 ratio. Here's how we would do it. We first enter our data. The easiest way is to enter our observed count data and then enter the expected *probabilities*.

```
> obs = c(50,20) # observed counts
> expP = c(0.75,0.25) # expected probabilities
```

Once we have our data entered, we do our test with the following command. You might notice that I haven't done a normality test. This test does not assume the data are normally distributed. Therefore, it is a non-parametric test.

```
> chisq.test(obs, p = expP)
```

```
        Chi-squared test for given probabilities

data:   obs
X-squared = 0.47619, df = 1, p-value = 0.4902
```

The chisq.test() function returns the chi-square statistic (χ^2 = 0.476), the degrees of freedom (df = 1), and the p-value (p = 0.49). The null hypothesis is that the observed values adhere to the 3 : 1 ratio (expP). In this example, we conclude that the data are consistent with a 3 : 1 ratio because p > 0.05.

We can understand this test best by looking at the equation used to calculate the χ^2 statistic:

$$\chi^2 = \sum_{i=1}^{n} \frac{(obs_i - exp_i)^2}{exp_i} \tag{10.1}$$

The obs_i and exp_i in the equation represent the count for the ith observed and expected values, respectively. In the 3:1 ratio example, we have just two categories (yellow and green). If the obs_i and exp_i are the same for all categories, then the χ^2 value is zero, which would be consistent with the H_0 that the observed data adhere to the expected. As the observed and expected values increasingly differ, the χ^2 statistic gets bigger. Once the χ^2 statistic reaches and/or exceeds a critical value, we find that $p \leq \alpha$ and we reject our H_0 that the expected and observed counts do not differ.

For the above example, where we had fifty and twenty individuals that should fall into a 3:1 ratio (52.5, 17.5), we can calculate χ^2 by hand using equation 10.1 as follows:

$$\chi^2 = \frac{(50 - 52.5)^2}{52.5} + \frac{(20 - 17.5)^2}{17.5} = 0.4762$$

I determined the expected values by multiplying 0.75 and 0.25 by the total number of individuals in the sample (70).

There are a few caveats with this test. First, you need to be careful about proportions versus counts (or frequencies). The above equation uses counts, while the function in **R** is best used with proportions. If you know it's a 3:1 ratio, for instance, then the expected frequencies should have 75% and 25% of the values, respectively. If you have 1371 observations, you'd have to calculate the number for each category ($1371 \cdot 0.75$ and $1371 \cdot 0.25$). No matter how many total observations you have, however, the proportions should still be 3:1, or 0.75:0.25.

A second concern involves limitations to the chi-square statistical test. The general rules are that there should be no more than 20% of the expected frequencies less than five and none of the expected frequencies should be less than one. If this is a problem, you should use the Fisher exact test (see section 10.3).

Data in a Contingency Table

You may encounter observed and expected data that have more than two categories. These may occur in a matrix, such as an n by m contingency table. The chi-square test can be used with these data. You need to enter your data into a matrix and send that matrix to the `chisq.test()` function. Let's try this with the following data. We might, for instance, have a sample with

Table 10.1 The number of males and females by hair color.

	Brown	Blond	Total
Males	10	6	16
Females	8	12	20
Total	18	18	36

males and females and two hair colors (brown and blond). If we assume that we have ten brown-haired males, six blond-haired males, eight brown-haired females, and twelve blond-haired females, we might be interested in whether individuals are distributed as expected across sex and hair color. This is an easy test to conduct but a really tricky test to interpret. The data are provided in table 10.2.

I've included the totals for the rows and columns in this matrix. The totals are used to calculate the expected values for the observed values. Notice that there are different numbers of males and females but the same number of blond- and brown-haired people in our population of 36 individuals. Given this, we can calculate the expected number of individuals that should occur in each category. For example, to get the expected value for brown-haired males, we multiply the row total by the column total and divide by the grand total:

```
> 16*18/36
```

```
[1] 8
```

We should have eight such individuals, but we saw ten, so we saw more than expected. Let's create the matrix of data and use **R** to do these calculations. We first enter the data into a matrix and verify they are entered correctly:

```
> mat = matrix(c(10,6,8,12), byrow = TRUE, nrow = 2)
> mat
```

```
     [,1] [,2]
[1,]   10    6
[2,]    8   12
```

Now we can do our statistical test by simply sending the matrix to the `chisq.test()` function:

```
> chisq.test(mat)
```

```
        Pearson's Chi-squared test with Yates' continuity
        correction

data:  mat
X-squared = 1.0125, df = 1, p-value = 0.3143
```

Table 10.2 Results from wart therapy.

Trmt	Resolved	Not
DT	22	4
Cryo	15	10

Note: number of subjects treated with duct tape (DT) or cryotherapy (Cryo); resolved or not resolved. Data from Focht et al. (2002).

Even though the observed frequencies differ from expected, they do not vary enough from what we would expect, given how many people fall into the two sexes and the two different hair colors ($\chi^2 = 1.012$, df = 1, p = 0.314). What if the statistical test is statistically significant? Let's look at another dataset.

10.2   An Example with Warts

Lots of kids get warts. Two-thirds of the warts go away on their own within two years. Kids think warts are gross and sometimes are willing to take extraordinary means to dig them out themselves. A common medical treatment is to freeze them with liquid nitrogen (cryotherapy). Another technique, tested by Focht, Spicer, and Fairchok (2002), was to use duct tape occlusion. The researchers tested the efficacy of placing duct tape over warts for two months. Fifty-one patients, who ranged in age from three to twenty-two years, completed the study. Each was randomly assigned to either the duct tape (DT) treatment or the cryotherapy (Cryo) treatment. Unfortunately, the researchers did not include a control group (do nothing and see what happens). The results are shown in table 10.2.

It is really hard to just look at these data and tell what happened. We do see, however, that most patients experienced warts being resolved and that it seems like a lot of patients who got the duct tape treatment lost their warts. The authors of the study conducted a χ^2 statistical test and concluded that duct tape significantly reduced the occurrence of the warts. Let's test this ourselves. We first need to enter the data:

```
> M = matrix(c(22,4,15,10),byrow = TRUE, nrow = 2)
> M
```

```
     [,1] [,2]
[1,]   22    4
[2,]   15   10
```

The data match the table. Let's visualize these data. The best way is to show the numbers of patients in each group using a barplot (see figure 10.1).

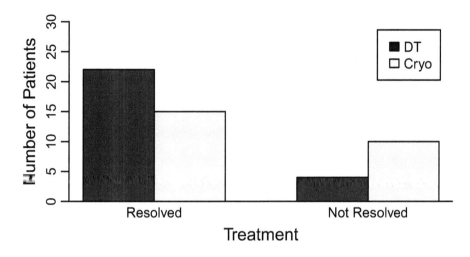

Figure 10.1 Barplot of the test for whether duct tape and cryotherapy reduced the prevalence of warts in patients.

```
> barplot(M, beside = TRUE, ylim = c(0,30),
+    xlab = "Treatment", ylab = "Number of Patients",
+    names = c("Resolved","Not Resolved"),
+    legend = c("DT","Cryo"), cex.lab = 1.5)
> abline(h=0)
```

To complete the χ^2 test, we simply need to send matrix M to the `chisq.test()` function:

```
> chisq.test(M)
```

```
        Pearson's Chi-squared test with Yates' continuity
        correction

data:  M
X-squared = 2.7401, df = 1, p-value = 0.09786
```

This is interesting. We get a p-value of 0.098. Why did the authors get a p-value of less than 0.05? The reason they got this answer is that they chose not to use the commonly used Yates's continuity correction. It turns out that **R**, by default, uses this correction due to the degrees of freedom in this contingency table being equal to one (df = 1). If we tell **R** not to do this correction, then what happens? Here we go:

```
> chisq.test(M, correct = F) # not using Yates' correction
```

```
        Pearson's Chi-squared test

data:  M
X-squared = 3.8776, df = 1, p-value = 0.04893
```

It is amusing that we have to tell **R** not to use the correct test. We now see how the researchers believed their results to be statistically significant ($p \leq 0.05$). What would you do if you were a doctor and had a patient with a wart? Based on these data there is no significant difference between using duct tape on warts versus using cryotherapy. Additionally, the researchers needed to include a control (applying no therapeutic treatment) to determine whether these treatments outperformed simply waiting for the warts to resolve.

Once we conduct our test correctly (with Yates's correction when we have a contingency table with df = 1) and determine that there is no statistical significance, we are done. We are not allowed to go on and ask, "Well, did those who received the duct tape occlusion treatment resolve more than expected?" Instead, we set up our experiment, we define our hypotheses, set our α, determine the experimental design and the statistical test we'll perform, conduct our experiment, collect our data, complete the *correct* statistical analysis, and interpret our result. Our correct results statement should read something like this:

> We found that the observed frequencies of patients in the two treatments (duct tape occlusion and cryotherapy) did not differ from what we would expect by chance among the patients tested ($\chi^2 = 2.74$, df = 1, p = 0.098).

10.3   The Fisher Exact Test

The previously discussed χ^2 test with the continuity correction is an approximation for this test (Zar, 2009), which uses, as the name implies, exact probabilities for observing frequencies. This test also allows us to employ one-tailed tests (the χ^2 test is only two-tailed). We can run this on a matrix much like we did using the chi-square test on the duct tape experiment.

```
> M = matrix(c(22,4,15,10),byrow = TRUE, nrow = 2)
> fisher.test(M)
```

```
            Fisher's Exact Test for Count Data

data:   M
p-value = 0.06438
alternative hypothesis: true odds ratio is not equal to 1
95 percent confidence interval:
  0.8333154 18.6370764
sample estimates:
odds ratio
   3.57229
```

This test on these data suggest what we had discovered from the chi-square test with Yates's correction: that the results are not statistically significant. Therefore, we conclude that using duct tape did not have an effect on the resolution of warts compared to the cryotherapy treatment (OR = 3.57, p = 0.064). Note that OR refers to the odds ratio.

10.4   Problems

1. Corn kernels were counted in an ear of corn. There were 295 purple kernels and 86 yellow kernels. We expect these to occur in a 3:1 ratio.

   (a) Enter these data into two variables called obs and exp so that you can perform a χ^2 test on them.
   (b) Are these counts consistent with a 3:1 ratio of purple to yellow?
   (c) Provide a graph of observed and expected counts.
   (d) Provide a formal results statement with correct statistical support.

2. Gregor Mendel reported (1866; see Hartl and Fairbanks, 2007) finding pea plants in one dihybrid cross experiment with the following ratio:

Phenotype	Number
Round/yellow	315
Wrinkled/yellow	101
Round/green	108
Wrinkled/green	32

   (a) Provide code and output from a chi-square test for these data, testing whether they adhere to a 9:3:3:1 ratio (in that order in the table).
   (b) Provide a barplot of just the data.
   (c) Extract from **R** the expected values for each phenotype had the plants occurred in exactly a 9:3:3:1 ratio (round the values).
   (d) Provide the observed and expected values in a single barplot.
   (e) Provide a formal results statement with correct statistical support.

3. The following data represent the observed body mass index (BMI) values for a male population with 708 individuals. Also included are the expected percentages for this particular population.

BMI	< 18	≥ 18 but < 25	≥ 25 but < 30	≥ 30
Observed	57	330	279	42
Expected percentages	10	50	30	10

   (a) Are the number of people in the BMI categories consistent with the expected percentages for this population? Be sure to include your code and output.

(b) Provide a side-by-side barplot of these data. The trace variable should be observed/expected.

(c) Extract from **R** the expected BMI value for each age class.

(d) Provide a barplot of the observed and expected BMI values in a barplot.

(e) Provide a formal results statement with correct statistical support.

CHAPTER ELEVEN

A Few More Advanced Procedures

IN THIS CHAPTER, we're going to gain a variety of skills that broaden our ability to explore biological data and models more deeply. This includes writing our own functions that simplify our analyses, digging deeper into understanding our data (e.g., curve fitting with a little bit of calculus), and writing and solving differential equation models. In the next chapter, we'll explore the components required to write a few simulation models.

11.1 Writing Your Own Function

So far you have used a large number of functions in **R**. If you look at the index of this book under "R functions," you'll see a long list of functions we've used. Sometimes, however, you might want to write your own function. Functions are lines of code that are bundled together into a single, hopefully well-named call. Functions can receive arguments; perform actions, such as calculations or create graphs; and can return information.

There are many advantages to creating functions. For instance, functions are useful to write when

1. You need to complete a particular, well-defined task that someone else hasn't already written.
2. You need to do something many times, and it is easier to write the task once, then call it many times with something like a "for" loop (see chapter 12).
3. You are collaborating and want to share a rock-solid piece of code so that it's easy for someone else to use and difficult for them to misuse.
4. You want to be able to compartmentalize tasks such as getting data, checking it for errors, and then making a graph, all with just a single line of code

To get started, let's write a function that simply returns the arithmetic mean of an array. This isn't necessary because **R** provides us with such a function (mean()). But it is a simple task that will be good for demonstration. The arithmetic mean ($\bar{x}$) for a set of data can be written as follows:

$$\bar{x} = \frac{1}{n} \sum_{i=1}^{n} x_i \tag{11.1}$$

Let's begin with some simple data:

```
> dat = c(6,3,4,5,3,1)
```

To write a function, you need a unique name for the function, that's descriptive. Use the function called function(), and decide whether it receives arguments, performs actions, and/or returns information. Here's a simple function for calculating the arithmetic mean:

```
> my.mean = function (x) {
+    ans = sum(x)/length(x)
+    return (ans)
+ }
```

In the function definition above, the data that are passed to the function are assigned to the variable x. The values in x are used to calculate the mean, which is then assigned to the variable ans. Finally, the function returns this answer. The variables x and ans are local to the function and are created and, when the function finishes, are removed.

To make the function available for later use during your **R** session, you need to run it. Just highlight all the code of the function and run that. If it runs, without error, you *might* have a working function. We usually want to test a function extensively before we consider it complete. Let's test the function:

```
> mean(dat)
```

```
[1] 3.666667
```

```
> my.mean(dat)
```

```
[1] 3.666667
```

The my.mean() function works on these data without generating an error and returns a value that is the same as **R**'s mean() function.

Functions can be fun to write. A really good function, however, can be challenging to write. The above function my.mean() is nice, but it won't provide much help if the user sends it bad data. You can look at most of the code for the built-in mean() function by typing this at the console:

```
> mean.default
```

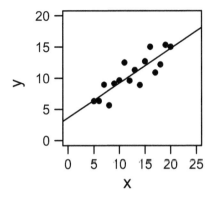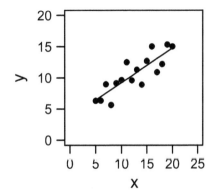

Figure 11.1 Two graphs with best-fit linear regression lines. On the left, I have incorrectly added a line with the abline() function that extends from axis to axis. In the right graph, I have correctly drawn a line with our new lm.line() function that extends the line only over the range of the x variable.

and hitting <enter>. You'll see a number of conditional terms (lines that start with if) as the author of the function (not me!) tests for a variety of ways that a user might make a mistake and therefore helps them out. A good function will check the incoming data and, if necessary, return an error message that's helpful (you've probably already seen several error messages provided by **R**).

Once you exit **R**, your own function will go away. For it to be available in your next **R** session, you'll have to run the function. Then it will become part of the **R** session and will be available and ready for you to use.

Let's build a function that allows us to add a proper best-fit line to a linear regression. We introduced this procedure in the section on linear regression (see section 9.2). To do this, we simply wrap our code in a function declaration and identify the arguments we want the user to be able to send to the function. Here's a simple version of the function:

```
> lm.line = function (x, y) {
+     lines (x, fitted (lm (y~x)))
+ }
```

I'm going to create some linear but noisy data and store the result from the linear regression analysis in a variable called mod.

```
> set.seed(100) # do this so you will have the same data
> x = 5:20
> y = 0.4*x + 2 + 7.5*runif (length (x))
> mod = lm (y~x)
```

I can now graph these data and compare the drawing of the line using abline() versus our own function lm.line() (see figure 11.1).

```
> par (mfrow = c (1,2)) # create a two-panel graphics window
> plot (x,y, xlim = c (0,25), ylim = c (0,20), pch=16,
```

```
+        cex.lab = 1.5, las = 1)
> abline(mod)
> plot(x,y, xlim = c(0,25), ylim = c(0,20), pch=16,
+        cex.lab = 1.5, las = 1)
> lm.line(x,y)
```

As you can see, we can now add appropriate best-fit lines to scatterplots using our lm.line() function. However, it's still important to make sure that all the assumptions have been met to add such a line correctly (see chapter 9).

11.2   Adding 95% Confidence Interval Lines for Linear Regression

When we conduct a regression analysis *and* get a significant result *and* the relationship between the *x* and *y* variables is linear, we can add a best-fit straight line. When we look at the analysis output, we see that our slope and intercept estimates have error terms. This is because the data points do not all fall exactly on the best-fit line. We can represent this uncertainty in regression with 95% confidence interval lines on either side of our best-fit line. This can help a reader interpret the strength of the regression relationship.

Let's start with a dataset that's built into **R** called "BOD," for biochemical oxygen demand. You can look at the BOD data by typing the dataset's name at the console. To add the 95% confidence interval lines, we need to conduct the linear regression analysis with the linear model function (lm()) and send the output from that function to the predict() function. This returns our 95% prediction values as a matrix that we send to the lines() function (see box 11.1). So we make the scatterplot, add the best-fit line, and then add the confidence lines (see figure 11.2).

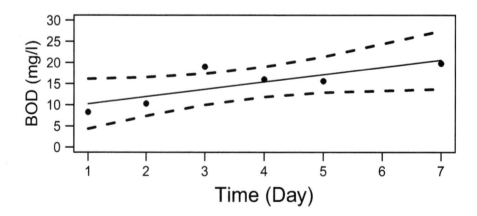

Figure 11.2 Scatterplot of the data with a best-fit linear regression line (solid line). In addition, the dashed 95% confidence lines for this relationship have been added.

```
> plot(BOD$Time, BOD$demand, pch = 16, ylim = c(0,30),
+       ylab = "BOD (mg/l)", xlab = "Time", las = 1,
+       cex.lab = 1.5)
> mod = lm(BOD$demand ~ BOD$Time)
> lm.line(BOD$Time, BOD$demand) # line function from earlier
> newx = BOD$Time # create a new variable for Time
> prd = predict(mod, interval = c("confidence"), level = 0.95,
+               type="response")
> lines(newx, prd[,2], lty=2, lwd = 2)
> lines(newx, prd[,3], lty=2, lwd = 2)
```

Box 11.1. *What do 95% confidence lines mean?*
This is a good question! They represent a confidence range, much like
the one we worked with when we calculated 95% confidence intervals and
placed them on barplots back in chapter 8. These lines show a range of
best-fit lines based on the variability in the estimates of both the intercept
and slope. This range is where 95% of the best-fit lines would occur if we
sampled data from the same population many times. These lines do not
capture 95% of the data points. They are intended to help the reader to
see how the variability in the data affect confidence in the best-fit line.
The dashed confidence interval lines in Figure 11.2 are fairly large, due
to the combination of variability in the data and there being relatively few
data points.

11.3   Nonlinear Regression

This section introduces more complicated models to fit to your data. If you
are thinking of fitting a curved line to your data, you should convince yourself
that your data satisfy the following criteria:

1. Your data clearly are not linear and can't be transformed to become linear.
2. The functional form that your data seem to adhere to can be represented
   by an equation that defines (not describes) the dynamics of the system.

Note that you are not fishing for some curvy line that looks good with
these data (recall the example of corals in this book's introduction)! You need
to understand how your data are related (e.g., how y depends biologically
on x) and fit that functional form to your data.

This would be a good time to consider the aphorism attributed to Albert
Einstein that a model should be made as simple as possible, but no simpler.
When we have data, we often are interested in finding the best model that fits

our data. That means we really want the equation that provides us parameter estimates of the model that governs how our system actually functions. Just because some model fits the data well is not confirmation that we've learned anything about our system! Fortunately, nature often is relatively simple and, therefore, a simple model is likely to be our best approximation about how our system functions. It is highly unlikely any biological system operates on a fourth-order or higher polynomial.

An Example Using Dolphin Mass over Time

A quick example should help (see figure 11.3). If we're interested in getting the functional relationship of y on x, then we should use the function that makes the most sense biologically (represents reasonably how the system changes over time). Below are mass data (kg) of a stranded dolphin over a ten-week rehabilitation period.

```
[1] 35.2 35.6 49.9 54.2 53.5 59.2 68.2 69.7 78.7 82.6
```

What is the rate of mass increase for our dolphin? Mammals, in general, are determinant growers, meaning they stop growing at adulthood. We can find out that bottlenose dolphins (*Tursiops truncatus*) can grow up to 300 kg. The masses above must represent an immature dolphin. The simplest model to describe the rate of change in the mass of our recovering dolphin would be to assume it increases linearly. We don't want to try to fit a high-order polynomial to our data just so the line goes through several (or every!) data point. A linear growth rate over this relatively short, ten-week period seems reasonable (see the left panel in figure 11.3).

Just for fun, I fit a ninth-order polynomial equation to these data (see the right panel in figure 11.3). Isn't this a lot better? This line goes through every data point exactly! If I use this model to predict the mass of this dolphin in the eleventh week, however, the model says it will weigh less than -1.5 billion kg. Funny, but not a very useful prediction. This model gives us neither an understanding of growth over time nor the ability to predict future mass of our dolphin. Choosing our model carefully is very important. Below is the code that creates figure 11.3.

```
> mod1 = lm(y~x)
> mod2 = lm(y ~ poly(x,9))
> par(mfrow = c(1,2))
> plot(x,y, xlim = c(0,11), ylim = c(0,100),
+       ylab = "Dolphin Mass (kg)", xlab = "Time (weeks)",
+       pch = 16, cex.lab = 1.5, las = 1, xaxt = "n")
> axis(1, at = seq(0,11,by = 5))
> lines(x,fitted(mod1),lwd = 2)
> plot(x,y, xlim = c(0,11), ylim = c(0,100),
+       ylab = "Dolphin Mass (kg)", xlab = "Time (weeks)",
+       pch = 16, cex.lab = 1.5, las = 1, xaxt = "n")
```

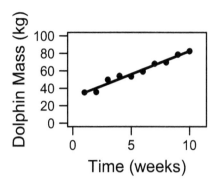 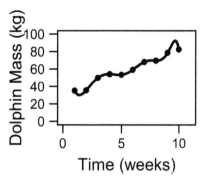

Figure 11.3 The change in average mass of a dolphin over ten weeks. In the left panel, we see the mass with a linear fit to the data (y = 5.3 week + 29; F = 253; df = 1,8; p < 0.001; adj. R^2 = 0.966). On the right, I fit a ninth-order polynomial equation. It's a great fit because the line goes through every data point. I'm not showing the equation because it's biologically meaningless!

```
> axis(1, at = seq(0,11,by = 5))
> xvals = seq(1,10,by=0.1)
> yvals = predict(mod2,list(x = xvals))
> lines(xvals,yvals,lwd = 2)
```

In summary, our linear model tells us that our dolphin is gaining a little over 5 kg per week (see figure 11.3).

Example: The Mass of Plants

Figure 11.4 shows the mass of different plants of a species grown in a green-house over the course of sixteen weeks. This measure of dry mass is destructive so each data point represents a different, independent plant. The species does not appear to grow linearly in mass over time. I have fit a straight line to the data, but we can see clearly that this line does not capture how these plants are growing. Note we don't just start fitting lines to data; we always use the correct functional form for our system. I'm adding different lines for demonstration purposes only!

```
> plant.mass = c(4.0, 8.5, 12.2, 13.4, 15.0, 17.8, 19.3, 19.4,
+        21.2, 21.7, 23.4, 23.8, 24.1, 24.7, 24.9, 25.5)
> time = 1:16 # weeks
> mod = lm(plant.mass ~ time)
> plot(time, plant.mass, xlim = c(0,16), ylim = c(0,30),
+      ylab = "Plant Mass (g)", xlab = "Time (weeks)",
+      pch = 16, cex.lab = 1.5, las = 1)
> lines(time,fitted(mod), lwd = 2)
```

Why isn't this a good model? It captures the basic increase of the data. However, we notice a clear pattern of the lack of fit. On both ends of the fit, the

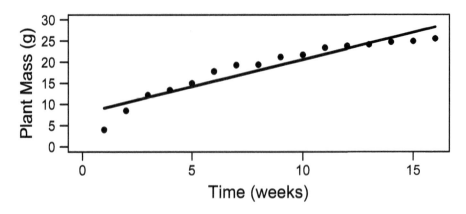

Figure 11.4 The mass of plants does not appear to increase in a linear fashion over time.

model overestimates the values of the data. Likewise, in the middle of the line, the model underestimates the data. This tells us that this model clearly is not describing the behavior of our system.

We are interested in the coefficients that describe the growth of plants but doing so will be more complicated than simply finding the best-fit straight line. We need to develop a model that properly defines the growth of these plants over time. A starting point for the mass of a plant at time zero is approximately zero. Therefore, our function should go through the origin.

How do we find the appropriate function and then get **R** to tell us what the equation is? Finding the right equation is hard. This usually comes from researchers with experience in the system and/or from previous work on plants. Once we have an equation, we then need to estimate the coefficients of the equation to help **R** get started searching for the best-fit coefficients. We must help **R** because the algorithm to find the best fit is numerical, meaning **R** uses a search strategy. It's quite possible **R** (like all other statistical programs) might begin searching in the wrong direction and never find the correct relationship. Because the data curve downward, let's try a simple quadratic function (a second-order polynomial). We can use the curve() function to approximate the relationship. I had to guess a bit to get started. For instance, the data curve downward, so the x^2 term must be negative.

After trying a few values for the coefficient, I got the curve function you see in figure 11.5. Below is the code that makes that figure, along with the equation, which is the first argument in the curve() function.

```
> plot(time, plant.mass, xlim = c(0,16), ylim = c(0,30),
+      ylab = "Plant Mass (g)", xlab = "Time (weeks)",
+      pch = 16, cex.lab = 1.5, las = 1)
> curve(-0.12*x^2+3.1*x+3,from = 1,to = 16, add = TRUE)
```

The line in figure 11.5 isn't a bad fit to the data and could be used to help **R** find the best-fit line. However, this function has a *very serious* problem.

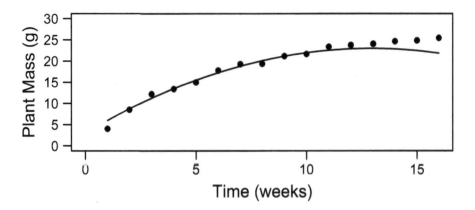

Figure 11.5 A second-order polynomial equation fit to the data. The line does a pretty good job of going through the data points, but can you see a problem lurking in the fit?

Can you see the problem? The curve fits the data pretty well, except near the end. That is really problematic! We're hoping to use an equation that describes the underlying mechanism that governs the growth of these plants. The equation that I have used fits the data well but completely misses *how* the plants grow. This equation suggests that the mass of these plants will get smaller and (wait for it) eventually be negative! Although we have done a good job approximating the equation for a line that fits these data, the underlying, quadratic model is terribly wrong! We will not use **R** to find the best-fit quadratic line through these data.

We need, instead, to find an equation that starts at the origin (seeds at time zero have approximately zero mass) and increases asymptotically. We can use a relatively simple, two-parameter model that goes through the origin and reaches an asymptote:

$$y = a \cdot (1 - e^{-bx})$$

For this function, we need to tell **R** some approximate values for the coefficients *a* and *b*. If we again look at the data, we might see that the asymptote seems to be around 25 grams dry weight. For our function, the asymptote is *a*. We can also see that if we make $x = 0$, then the intercept is zero. I don't know the value of *b*, but it controls the rate at which the line reaches the asymptote. That sounds like it's a *really* important growth rate parameter to estimate. Let's just start it at $b = 1$ and see what happens using the curve () function again (see figure 11.6).

```
> plot(time, plant.mass, xlim = c(0,16), ylim = c(0,30),
+       ylab = "Plant Mass (g)", xlab = "Time (weeks)",
+       pch = 16, cex.lab = 1.5, las = 1)
> curve(25 * (1 - exp(-(1*x))),from=0,to=16,add = TRUE)
```

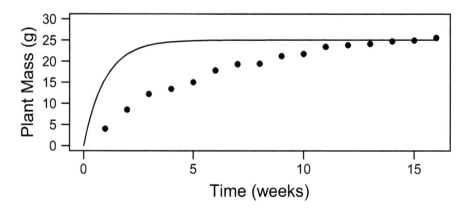

Figure 11.6 An asymptotic function that is a poor fit to the data but represents a relationship that seems to have the right idea.

We can see that our function gets to the asymptote too quickly. But it does have a form that includes an asymptote and is heading in the right direction. Let's ask **R** to approximate the best-fit values for our model:

```
> mod = nls(plant.mass ~ a*(1 - exp(-b * time)),
+           start = list(a = 25, b = 1))
```

Notice that exp() is the exponential function for the natural logarithm. The nls() function conducts the nonlinear, least-squares analysis. It needs the mathematical function we're trying to fit and our best-guess estimates for the coefficients. We need to send these starting values as a list.

If our calling of the function doesn't return an error, then our model results are stored in the variable fit. We can now send fit to the function summary() and see the parameter estimates and the estimates of variability for these:

```
> summary(mod)
```

```
Formula: plant.mass ~ a * (1 - exp(-b * time))

Parameters:
   Estimate Std. Error t value Pr(>|t|)
a 26.828212   0.439173   61.09  < 2e-16 ***
b  0.176628   0.007396   23.88 9.59e-13 ***
---
Signif. codes:
0 '***' 0.001 '**' 0.01 '*' 0.05 '.' 0.1 ' ' 1

Residual standard error: 0.5266 on 14 degrees of freedom

Number of iterations to convergence: 6
Achieved convergence tolerance: 7.27e-07
```

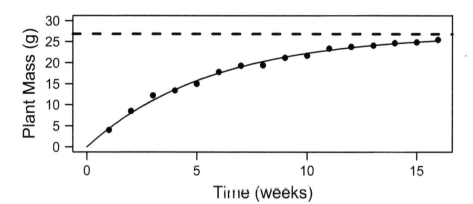

Figure 11.7 A best-fit asymptotic function for our data for the change in the mass of plants over time. The horizontal line above the curve is our asymptote.

From this, we can see estimates of the coefficients and error estimates for each coefficient. The equation is therefore:

$$mass = 26.8 \cdot (1 - e^{-0.177 \cdot week})$$

In the output, we also see that **R** has provided the p-values for each parameter estimate. In this example, we see that the estimate of each parameter is highly significant ($p < 0.001$). This is important to us because we need to verify that the coefficients we've asked **R** to fit with our equation are actually important. If we overfit our model (have too many parameters) we would find nonsignificant coefficients. This would suggest that we should consider a simpler model.

This equation is exciting because it gives us some insight about how our plant changes in mass over time. We also can compare our results with results by other researchers. We can test whether the parameters change given different environmental conditions. And we can ask what the instantaneous rate of change is using calculus. All sorts of great things become possible! Here's the final graph with the correct best-fit line (figure 11.7).

```
> plot(time, plant.mass, xlim = c(0,16), ylim = c(0,30),
+    pch = 16, ylab = "Plant Mass (g)", xlab = "Time (weeks)",
+       cex.lab = 1.5, las = 1)
> xv = seq(0,16,0.1)
> yv = predict(mod, list(time = xv))
> lines(xv, yv)
> asymptote = sum(coef(mod)[1]) # asymptote is the
>       # sum of these two coefficients
> abline(h = asymptote, lwd = 2, lty = 2)
```

In the code above, I have captured the value of the asymptote by using the coef() function, which extracts just the estimated values of the coefficients from the model. We store that result in the variable called, not surprisingly, `asymptote`. We then send that calculated version to the `abline()` function and have it add a horizontal reference line (hence the `h = asymptote` argument) to our graph. I've asked for a red line, although you're probably seeing the line in the figure in black.

Get and Use the Derivative

In biology, we often have situations where something is changing over time or as a function of some other variable. We graph our points and wish to fit a function to the data to learn more about what is going on or to make some prediction. One such problem commonly encountered is with the Michaelis-Menton relationship. Many introductory laboratories include a module on looking at enzyme kinetic reactions as a function of substrate concentration.

Here are the steps, or the algorithm, for finding the rate at any point along a Michaelis-Menton function derived from data:

1. Get the data into two arrays of equal length.
2. Create a scatterplot of the data.
3. Find the best-fit function for these data using the Michaelis-Menton equation v = Vmax*[S]/(Km + [S]).
4. Add the best-fit line to the graph.
5. Get the derivative of the Michaelis-Menton function.
6. Find the slope of the tangent at a give value of [S].
7. Add the tangent line to the best-fit curve for the data.
8. Report the slope of the tangent line (the rate) and this given value of [S].

Here are some sample data for the substrate concentration (S) and the velocity of the reaction (v). They are graphed in figure 11.8.

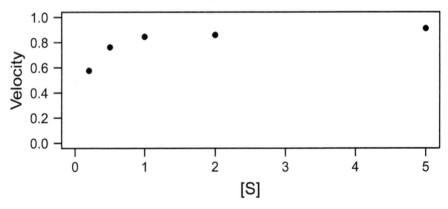

Figure 11.8 A scatterplot of the data for velocity and substrate concentration.

```
> S = c(0.2,0.5,1,2,5) # x-axis data ([S])
> v = c(0.576, 0.762, 0.846, 0.860, 0.911) # velocity

> plot(S,v, xlab = "[S]", ylab = "Velocity", xlim = c(0,max(S)),
+       ylim = c(0,1), cex.lab = 1.5, pch=16, las = 1)
```

To determine the best-fit line, we use the nonlinear, least-squares function (nls()), which returns the coefficients for the model. However, we must tell **R** what model to use. In this case, we'll specify the Michaelis-Menton function $\left(v = \frac{V_{max} \cdot S}{K_m + S} \right)$. As we have seen, **R** requires us to provide educated guesses of the coefficients we're fitting. Note that our unknowns are V_{max} and K_m. **R** will use the starting values we provide to begin its search for the best parameter estimates.

```
> mod = nls(v ~ (Vmax * S)/(Km + S),
+           start = list(Vmax = 1, Km = .1))
```

If it works without returning an error, our resulting model will be stored in the variable mod. We can now add the best-fit line that **R** has given us to the data in our graph. We can now send mod to the summary() function to see the result:

```
> summary(mod)
```

```
Formula: v ~ (Vmax * S)/(Km + S)

Parameters:
      Estimate Std. Error t value Pr(>|t|)
Vmax 0.930469   0.011879   78.33 4.59e-06 ***
Km   0.118509   0.009646   12.29  0.00116 **
---
Signif. codes:
0 '***' 0.001 '**' 0.01 '*' 0.05 '.' 0.1 ' ' 1

Residual standard error: 0.01536 on 3 degrees of freedom

Number of iterations to convergence: 4
Achieved convergence tolerance: 4.272e-07
```

We can see that the two parameters have estimates and variabilities (standard errors). Let's use the model output to draw the best-fit line onto the graph with the data (see figure 11.9). We use the predict() function to give us the model's y-values and draw that line over the range of x-values.

```
> plot(S,v, xlab = "[S]", ylab = "Velocity", xlim = c(0,max(S)),
+       ylim = c(0,1),cex.lab = 1.5, pch = 16, las = 1)
> x = seq(0,max(S),by = 0.01) # an array seq. of x values
> y = predict(mod,list(S = x)) # the predicted y values
> lines(x,y, lwd = 2) # draw the line
> asymptote = coef(mod)[1] # value of the asymptote
> abline(h = asymptote)
```

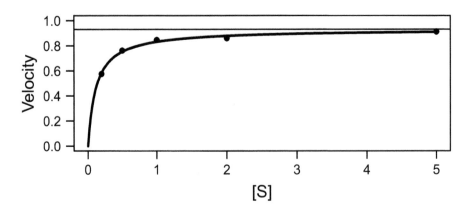

Figure 11.9 A scatterplot of the data for velocity and substrate concentration with the best-fit line. The equation for the line, returned by the `nls()` function, is $v = \frac{0.930 \cdot S}{0.118 + S}$.

Get the Slope of the Tangent Line

We can get the slope at any point on our function by taking the derivative of our function. As we have seen above, the coefficients are saved in the variable "mod" so we can get these parameter estimates individually. Let's get them from the mod and just call them a and b for simplicity. To do this, we can use the function `coef()`.

```
> Vmax = coef(mod)[1] # this is Vmax
> Km = coef(mod)[2] # this is Km
```

Now we need the derivative to the Michaelis-Menton equation. We tell **R** what the equation is and then get the derivative with the function `D()`.

```
> my.exp = expression(Vmax * my.S/(Km+my.S))
> my.deriv = D(my.exp,"my.S")
> my.deriv # look a the derivative returned by D()
```

```
Vmax/(Km + my.S) - Vmax * my.S/(Km + my.S)^2
```

I'm going to choose a value of [S] so that the tangent line intercepts my function. I'll call that particular value my.S. I don't want to use the original variable "S" because that will overwrite the original data.

```
> my.S = 1 # choose where on [S] to get the rate (tangent line)
```

The next step we need to take is to find the tangent line at [S] = my.S. To do this, we need to find the slope at a particular value of [S] at my.S = 1. We can also add a vertical line at this point to help convince ourselves we're finding the tangent at this point along our function. The following code combines these elements to create figure 11.10.

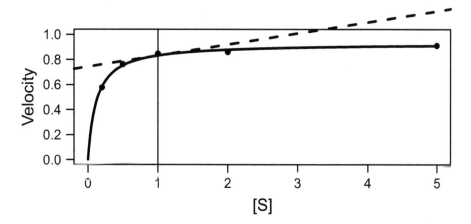

Figure 11.10 A scatterplot of the data for velocity and substrate concentration with the best-fit line. The equation for the line, returned by the nls() function, is $v = \frac{0.930 \cdot S}{0.118 + S}$ with a tangent line added at [S] = 1.

```
> plot(S,v, xlab = "[S]", ylab = "Velocity", xlim = c(0,max(S)),
+       ylim = c(0,1), cex.lab = 1.5, pch = 16, las = 1)
> der.slope = eval(my.deriv) # get slow using deriv. at my.S
> der.y = eval(my.exp) # Ht of the best-fit function at my.S
> der.int = der.y - der.slope*my.S # The intercept of tangent
> lines(x,y, lwd = 2)
> asymptote = Vmax
> abline(der.int,der.slope,lwd = 2,lty = 2) # draw tangent
> abline(v=my.S) # place a vert line at tangent
```

Finally, let's have **R** report the slope of the tangent at this point for our chosen [S].

```
> cat("The velocity of the reaction at",
+      my.S," is ",der.slope,"\n")
```

```
The velocity of the reaction at 1  is  0.08814004
```

I suggest you try to draw that tangent line in another place and get the velocity at that substrate concentration level.

11.4   An Introduction to Mathematical Modeling

In addition to being able to conduct statistical tests, we sometimes need to create and evaluate mathematical models. For instance, we might be interested in using the statistical fit from data and then use those parameter estimates to build a model, such as exponential growth for a population. We'll look at this model and then at a more complicated, standard epidemiology model.

Exponential Growth

Exponential growth is a relatively simple differential equation model:

$$\frac{dN}{dt} = rN \tag{11.2}$$

This model defines the instantaneous rate of change of the population (dN/dt) as a function of a growth rate parameter (r) times the population size N. This equation defines the rate of change of the population. But what is the population size at any given time? To determine this we need to solve this differential equation, which means we need to integrate it. If we integrate this model we find the solution:

$$N_t = N_0 e^{rt} \tag{11.3}$$

In practice we are unable to solve most differential equations of biological systems analytically to find their solutions. Instead, we have to rely on programs like **R** to *approximate* the solution for us *numerically*. Let's solve this relatively simple differential equation model numerically to see how this is done in **R**. We're going to use some code that is admittedly challenging to understand. We'd love to take time to figure this out, but we're going to just implement the approach here. Through this example and the next, you'll be able to solve a variety of other problems with minor changes to this code.

Instead of using the analytical solution (equation 11.3) we'll ask **R** to approximate the solution from equation 11.2. Below we define our constants, preferably from data, and then define our model. You may need to install the deSolve package.

```
> library(deSolve)  # library for solving diff. equations
> num_yrs = 10
> r = 0.2 # the growth rate parameter
> N0 = 100 # starting population
> xstart = c(N=N0) # create a "list" of starting values
> parms = r # here's our one parameter
> mod = function(t,x,parms) {
+    N = x[1]
+    with(as.list(parms) , {
+      dN.dt = r*N
+      res=c(dN.dt)
+      return(list(res))
+    })}
```

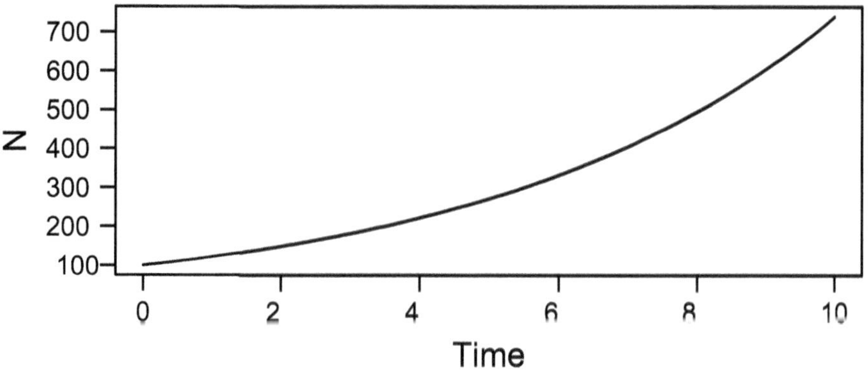

Figure 11.11 The graphed solution to exponential growth (equation 11.2).

Once we have run the model, defined in our function called mod, we send it, along with the parameter r and starting values (N0), to the ode () function. The results will be stored in the variable output.

```
> time=seq(0,num_yrs, by = 0.1)   # set number of time steps
> # RUN THE MODEL in the next line!
> output = as.data.frame(ode(xstart,time,mod,parms))
```

Finally, you should look at what's stored in the dataframe output. We can plot those results as follows (see figure 11.11):

```
> plot(output$time,output$N, xlab = "Time", ylab = "N",
+       type = "l", cex.lab = 1.5, las = 1)
```

The SIR Model

In our second model, we will see the outcome from a standard epidemiology model, called a susceptible, infectious and recovered individuals (SIR) model. This model is the starting place for understanding the dynamics of diseases such as influenza or COVID-19. Here's the model, called a system of differential equations by mathematicians and a coupled differential equation model by biologists. The bottom line is that these equations depend on each other because they share state variables (S and I).

$$\frac{dS}{dt} = -\beta SI$$

$$\frac{dI}{dt} = \beta SI - \nu I \qquad (11.4)$$

$$\frac{dR}{dt} = \nu I$$

Each equation governs the dynamics of each of these groups of individuals. The model is technically closed because no individuals enter or leave the model (they only can move from $S \to I \to R$). Susceptible people that come into contact with infectious individuals become infectious and leave the S class at a rate controlled by the coefficient β. Thus, the term for $\frac{dS}{dt}$ is negative ($-\beta SI$). Infectious individuals recover at rate ν.

The system of equations (equation 11.4) is clearly intimidating to biologists. But it is easy to solve using **R**. What biologists want to know is how the prevalence of a disease might change over time. In particular, if you are interested in the health sciences, you might be working to making diseases disappear. With the COVID-19 pandemic, we hear a lot about flattening the curve. The guidelines for wearing masks and practicing physical distancing are efforts to reduce β. If we measure infection and recovery rates, we might find that we get the dynamics seen in figure 11.12.

```
> Num_Days = 20   # number of days to run simulation
> B = 0.006       # transmission rate (0.006)
> v = 0.3         # recovery rate (0.3)
> So = 499            # initial susceptible pop (299)
> Io = 1              # initial infectious pop (1)
> Ro = 0              # initial recovered pop (0)
> xstart = c(S=So,I = Io, R = Ro)
> parms = c(B, v)
> times=seq(0,Num_Days,length=200)   # set up time steps
```

The code below defines the model as a function. The first part sets the initial values for S, I, and R from the initial values sent to the function as the x argument. The with() function contains the actual mathematical model (equation 11.4).

```
> mod = function(t,x,parms) {
+   S = x[1] # init num of susceptibles
+   I = x[2] # init num of infectious
+   R = x[3] # init num of recovered
+   with(as.list(parms) , {
+     dS = -B*S*I       # dS/dt
+     dI = B*S*I - v*I  # dI/dt
+     dR = v*I          # dR/dt
+     res=c(dS,dI,dR)
+     list(res)
+   })}
```

We can now run the model, much like we did for the exponential growth model. The actual running of the model is done with the following line of code:

```
> output = as.data.frame(lsoda(xstart,times,mod,parms))
```

It seems pretty simple, but what **R** is doing behind the scenes is really amazing (and complicated)! It is approximating the solution to the system of differential equations and storing the result in the dataframe output.

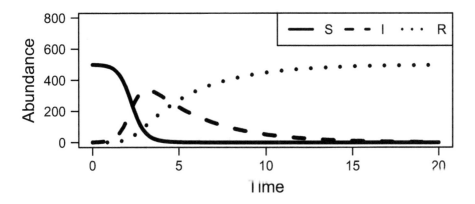

Figure 11.12 Output from the SIR model.

The remaining code creates the plot and draws the S, I, and R populations using the lines() function.

The last step is to make a presentation-ready visualization. In the code below, we do something odd: we call the plot function but then use "type = "n"," which means the graph is made but we don't add anything to it. We then add three lines to the set of axes for the susceptible, infectious, and recovered individuals, and then end with adding a legend (see figure 11.12).

```
> plot(output$time, output$S, type="n",ylab="Abundance",
+      xlab="Time", main="ODE Model", cex.lab = 1.5,
+      ylim = c(0,So*1.6), las = 1)
> lines(output$time,output$S,lty = 1,lwd=3)
> lines(output$time,output$I,lty = 2,lwd=3)
> lines(output$time,output$R,lty = 3,lwd=3)
> leg.txt = c("S","I","R")
> legend("topright",leg.txt,lwd=2,lty = 1:3)
```

11.5   Problems

1. Write a function called my.stats() that takes an array of numbers and gathers the mean, standard deviation, and standard error of the mean into a dataframe and returns it. Test your function with data and verify that your function returns the correct values.

2. Using the built-in Loblolly pine dataset, graph the change in the height of trees of seed type 301. These trees seem to exhibit linear growth over time. If the relationship is significant, add the best-fit line. Include 95% confidence interval lines to the graph.

3. Below are data for my motorcycle's gas mileage in miles per gallon (mpg) on level ground, determined instantaneously at different speeds in miles per hour.

mpg	44	52	58	63	58
Speed	20	30	40	50	60

Assume that the function that defines these data is a second-order quadratic function (e.g., $y = a \cdot x^2 + b \cdot x + c$).

(a) Create a publication-quality graph of these data with the best-fit quadratic function added to the data.

(b) What is the optimal speed for my motorcycle to travel in order to maximize my mileage (miles per gallon)? Note that you need to find the underlying function for mpg as a function of speed for my motorcycle. The maximum mpg is clearly 50 mph in these data. However, the question is, What is the best speed based on these data? You need to find the most appropriate curve for these data (it is well approximated by a second order polynomial polynomial) and then find the speed where that curve has a slope = 0. This is best solved using calculus.

4. Consider the following logistic growth function:

$$\frac{dN}{dt} = rN(1 - \frac{N}{K}) \tag{11.5}$$

which has the following analytical solution:

$$N_t = \frac{K \cdot N_0 \cdot e^{r \cdot t}}{K + N_0 \cdot (e^{r \cdot t} - 1)} \tag{11.6}$$

where the initial population is $N_0 = 50$, $r = 0.82$, and $K = 1000$.

(a) Use **R**'s solver (deSolve) to find and graph the solution to the differential equation (11.5) over the time period $0 \le t \le 10$.

(b) Use the curve () function to plot the solution (equation 11.6) from $t = 0 - 10$.

5. Enzyme kinetics can be described with the following Michaelis-Menton equation:

$$v = \frac{V_{max} \cdot [S]}{K_m + [S]}$$

where v is the velocity of the reaction, V_{max} is the maximum velocity, $[S]$ is the concentration of the substrate, and K_m is the Michaelis-Menton constant.

(a) The built-in dataset called `Puromycin` contains data on the reaction rate versus substrate concentration for cells treated and
untreated with Puromycin. Create side-by-side scatterplots for
the "treated" and "untreated" data. *Hint:* You will need to
`subset()` the data.

(b) Add to each scatterplot the best-fit relationships.

CHAPTER TWELVE

An Introduction to Computer Programming

IN ADDITION to the statistical, mathematical, and visual power of **R**, you will find that **R** also is a full-featured, object-oriented programming language. In general, we use programming to get the computer to perform tasks that need to be done many times. Our problems can be quite complex, so we also need to control the problem-solving route to include conditional statements, such as "If this is true, then do task 1, otherwise do task 2." This is done a lot in biology for a variety of problems. This introduction is meant to give you a taste of what is possible. Knowing what's possible can allow you to think of new problems that you might consider solving.

12.1   What Is a Computer Program?

Computers need to be told exactly what you want them to do. And they are really good at doing what we tell them to do (like when it's time to show a user the so-called blue screen of death). And they're particularly good at doing things repetitively.

For our programming needs, computers run sets of instructions. These instructions are written by people like you and me that use a high-level language that makes sense to us. Other programs, written by other humans, then take the instructions and convert them into low-level instructions that computers understand. My goal here is to introduce you to the high-level language of **R** so you can solve problems in biology.

There are many high-level languages that can be used to create computer programs. In addition to **R**, you may have heard of "Matlab," "Mathematica," "Visual Basic" (used within Excel), and "Python." Different languages have their own strengths and weaknesses. Strangely, there is no *best* language. For modeling complex biological systems, however, most any problem can be solved using **R**.

The advantage of **R** is that there is a very large set of easy-to-use, built-in tools for creating graphs, running statistical tests, and analyzing mathematical models. Although **R** programs run relatively slowly, the development time for writing programs in **R** is very short. Nonetheless, it really helps if you know what you're trying to do before you begin coding. As you develop your skills in programming, it's good to know that, if you can program in one language, it becomes relatively easy to migrate to a new language.

This chapter is just an introduction, so we won't be tackling a complex problem in biology. You will have to continue to explore and read about your problem and programming. I hope that you can get started, however, and see why you might want to do this and how programming can be used to understand and predict patterns in biological systems.

Computers are great at doing lots of calculations that we'd find pretty boring to do by hand. To accomplish these tasks, we use just a few basic constructs when building programs:

1. if (else) statements (conditional tests)
2. for loops (control flow)
3. while loops (control flow)
4. functions that combine instructions into a single call

The if statement is used to test whether a condition is true. If the condition evaluates to be true, then the following statement is executed. If the conditional test is false, then the true expression is ignored. You can provide an expression to execute if the statement is false with an else expression.

We use for and while loops when we want a set of instructions executed more than once. The for loop should be used if you know how many times you want to do something (we usually know this). If the number of times some instructions need to be done is determined during run time, then you use a while loop. The while loop is done until some criterion is satisfied.

An Example: Modeling the Central Limit Theorem (CLT)

Let's build a short program that tests the central limit theorem (CLT), which states that the distribution of sample means from any distribution is approximately normal. To accomplish, this let's take 5,000 samples (not values) from the standard normal distribution ($\bar{x} = 0$ and $\sigma = 1$) and a uniform distribution ($x \in [0, 1]$). Because we know how many times we want to do the sampling, it is most appropriate to use a for loop.

To test the CLT, we will perform a simulation that requires us to use a random number generator. Let's begin by setting the random seed so we'll get the same results:

```
> set.seed(20) # do this so you will have the same data
```

We can then declare our sample size (num.values) and the number of times we take samples from each population (num.samples), as follows:

```
> num.values = 100 # the number of observations per sample
> num.samples = 5000 # the number of samples to draw
```

For each of the 5,000 samples, we need to calculate and store the mean of the 100 values from each sample. Therefore, we need two structures that will each hold a list of 5,000 mean values. For this, we'll use arrays. Arrays can be declared as follows:

```
> means.norm = numeric(num.samples) # will hold 5000 values
> means.unif = numeric(num.samples)
```

At last we are ready to perform our simulation and collect our data.

```
> for (i in 1:num.samples) {
+    means.norm[i] = mean(rnorm(num.values))
+    means.unif[i] = mean(runif(num.values))
+ }
```

The above code does all the work. It tells **R** to run the two lines between the curly braces ({ }) 5,000 times (from 1 to num.samples). The first line within the for loop works like this:

1. Send num.values (100) to the rnorm() function.
2. The rnorm() function returns 100 random numbers drawn from the standard normal distribution ($\bar{x} = 0$, $s = 1$).
3. Those 100 values from the standard normal distribution are then sent to the mean() function.
4. The mean() function calculates the arithmetic mean of those numbers and assigns that result to the ith element in the means.norm[i] array.

Once the for loop is done, the variables means.norm and means.unif should hold 5,000 values each. We now can look at the distributions of these means using the hist() function (see section 5.2 on histograms). I've also plotted the distributions of single samples from the standard normal and uniform distributions for comparison (see figure 12.1).

```
> par(mfrow = c(2,2))
> hist(rnorm(num.samples),xlab = "X",main = "Normal Sample",
+      ylim = c(0,1000), las = 1)
> hist(runif(num.samples),xlab = "X",main = "Uniform Sample",
+      ylim = c(0,1000), las = 1)
> hist(means.norm, xlab = "X",main = "Means from Normal Dist",
+      ylim = c(0,1000), las = 1)
> hist(means.unif,xlab = "X",main = "Means from Uniform Dist",
+      ylim = c(0,1000), las = 1)
> par(mfrow = c(1,1))
```

The histograms in the lower part of figure 12.1 appear similarly bell-shaped. Note that the histogram in the lower right is from 5,000 means

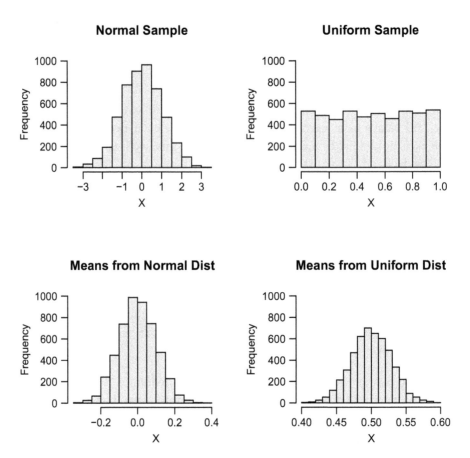

Figure 12.1 A visualization of the central limit theorem (CLT). The upper panels show the distribution of two samples of 5,000 values. The sample on the top left is drawn from a normal distribution and the sample on the top-right is drawn from a uniform distribution. The bottom panels demonstrate that the means of many samples drawn from a normal (lower-left panel) and uniform (lower-right panel) distribution result in relatively normal distributions, in agreement with CLT.

of samples drawn from the far-from-normal, uniform distribution. We can test the distributions of means from these two distributions for normality using the Shapiro-Wilk test (see section 4.4):

```
> shapiro.test(means.norm)$p # from normal pop
```

```
[1] 0.719782
```

```
> shapiro.test(means.unif)$p # from unif pop
```

```
[1] 0.3899854
```

What do we conclude about these distributions? Does this result agree with or contradict the central limit theorem?

What we did above was to write a computer program that completed a task 5,000 times in the blink of an eye. As we discussed above, we sometimes want to have a computer test an idea that might require many calculations. This is a great example that has allowed us to verify empirically (that means actually doing it) a theorem. Our next step is to explore further how programming can be used to answer some questions in the biological sciences.

12.2   Introducing Algorithms

Let's approach our introduction to algorithms with an example of simulating genetic drift operating on two alleles found at a single locus. Genetic drift is a mechanism that brings about evolution in a population, or the change in gene frequencies over time. Genetic drift is a random process, unlike natural selection. The process is governed by chance: one allele might become more or less frequent over time, simply by chance.

For this exercise, let's assume that individuals are haploid: they have either a 0 or 1 allele at a single locus. This way, the proportion of alleles that are of type 1 will simply be the sum of the allele values divided by the number of individuals. If we had a small population with six individuals and each has the alleles 0, 1, 1, 0, 0, 0, the proportion of the 1 allele is just 2/6 = 0.333.

We'll assume that reproduction is really simplistic. The population size will remain the same over time and that reproduction is completed by randomly choosing N individuals from the current population. The chosen individuals each make a baby that is identical to itself and is placed into the next generation.

What *should* happen over time? We might, just by chance, select more individuals with the 0 allele than the 1 allele over time and end up with a population of only the 0 alleles (this is called fixation in population genetics). It is just as likely that we might end up with the 1 allele becoming the only allele in the population. To build our model, we begin by developing an algorithm, which is somewhat similar to a recipe that guides us in writing our program to simulate this process. Here are the steps that I follow to complete this simulation (there are many other ways you could do this):

1. Declare the necessary variables (e.g., N).
2. Create an empty plot for our results.
3. Run the simulation many times (sounds like a "for" loop). In each simulation we:

    (a) Create a population.

(b) Store the proportion of 1s in a variable (e.g., *P*) for each time step.

(c) Perform the reproduction routine, randomly choosing individuals that will contribute to the next generation and do this many times.

(d) Add a line for the P data to graph.

(e) If not done, repeat for the next replicate.

We now need to implement these steps, assuming that the algorithm will work. What I do is write the algorithm in my script file and then write the code directly below each task. I can then test each task individually to make sure it's doing what I think it should. If the tasks are more complicated, I might write a function to do the work. The code below creates figure 12.2. Note that the first line calls the set.seed() function (with 10 as the argument) so that if you run this code, you should get exactly the same graph as I did. (Remember, don't include the ">" and "+" symbols in your script file.)

```
> set.seed(10) # do this so you will have the same data
> # Step 1: Define variables
> n.time.steps = 100 # how long to run simulation
> pop.size = 50 # how many individuals there are
> n.reps = 20 # how many populations to simulate
> # Step 2: Create an empty plot for simulation
> plot(0,ylim = c(0,1),xlim = c(1,n.time.steps), type = "n",
+         ylab = "P(Allele 1)",
+         xlab = "Time Step", cex.lab = 1.5, las = 1)
> abline(h=0.5,lty = 2, lwd = 3)
> # Step 3: Run the simulation n.reps times
> for (i in 1:n.reps) {
+    # Step 3a: Create population of 0 and 1 alleles
+    pop = c(rep(0,pop.size/2),rep(1,pop.size/2))
+    # Step 3b: Store proportion of 1s in variable
+    P = sum(pop)/pop.size
+    # Step 3c: Run the simulation for this replicate
+        for (j in 2:n.time.steps) {
+                pop = sample(pop,pop.size, replace = T)
+                P[j] = sum(pop)/pop.size
+        }
+    # Step 3d: Add line to graph for this replicate
+    lines(P)
+ }
```

This simple model has been used to make a variety of important contributions to the fields of population genetics and conservation biology. It also is an excellent start on an undergraduate research project.

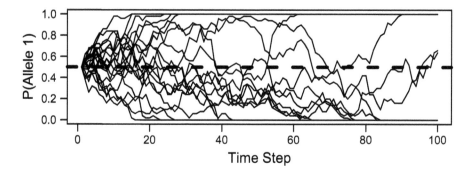

Figure 12.2 The proportion of allele 1 in twenty replicate populations over time. Each line represents a separate population. Eighteen of the twenty replicates result in either the loss or fixation of allele 1. Two simulations continue, with both alleles persisting in these populations.

12.3   Combining Programming and Computer Output

Sometimes we would like to have code that solves a problem and provides the answer to whomever is looking at the screen (ourselves or others). We can use commands that will provide the answer with some regular English mixed in. This can be great to help out lab mates, for instance. Here's how we might solve a small problem with a contingent result and print the appropriate result to the console.

Imagine that you want to provide your lab group with a chunk of **R** code that will take their data and tell them whether the data are or are not normally distributed. Here's a way to do this using a function that outputs the answer using the cat() function.

```
> normality.test = function(x) {
+    ans = shapiro.test(x)
+    if(ans$p.value > 0.05) {
+        cat("The data are normally distributed:
+        p = ",ans$p.value)
+    } else {
+        cat("The data are not normally distributed:
+        p = ",ans$p.value)
+    }
+ }
```

Your lab group simply needs to run the code above and enter their data into the x variable below and run the function call. The function call is a more intuitive name.

```
> x = c(1,2,5,3,2,1) # enter your data into this variable
> normality.test(x)
```

```
The data are normally distributed: p =   0.2117055
```

At this point in your statistics training with **R**, you might think this is pretty simple. It hardly does anything more than the simple Shapiro-Wilk test does for us. However, some folks haven't read this book and aren't sure what the p-value means, especially in this particular statistical test. So this code can be really helpful by writing out the answer.

The tools we have used above represent really important components for creating programs in **R**. With these tools, and perhaps a few more that you'll pick up as you go, you'll be able to answer a variety of really interesting questions in biology. Test yourself by trying the problems in the next section. Have fun! The key to success in computer programming is play. If you're interested in doing more programming, check out the resources listed in the Introduction and by doing a Google search. One good, complete guide is the book by Matloff (2011).

12.4   Problems

1. The gamma distribution can result in samples that are highly skewed to the right. Test whether the means of 5,000 samples, each of size 100 values, drawn from the gamma distribution, are normally distributed (see figure 12.1). Use the following sample code:

```
> rgamma(100, shape = .5)
```

2. Create a program that uses a `for` loop to count the integers from 1 to 100.

3. The Fibonacci series begins with the numbers 0, 1, and continues as the sum of the two previous numbers (0, 1, 1, 2, 3, 5, ...).

   (a) Write an algorithm that solves this problem.
   (b) Write a program that generates the first twenty numbers of the series.
   (c) Create side-by-side graphs (in a single graphics window) of this series. On the left, present the series on a linear scale; on the right, present it on a semi-log (ln) scale (y is logged, x is linear).
   (d) Write a function called `Fibonacci()` that takes as a single argument the number of Fibonacci values to print out. It should assume that the starting values are 0 and 1. The function should return the series with the correct number of values.

4. Zebra mussels (*Dreissena polymorpha*) are invasive, freshwater gastropods that are expanding their range in North America. When individuals invade a lake, their population is capable of growing rapidly. Assume that

a newly introduced population in a lake, grows geometrically, according to the following difference equation:

$$N_{t+1} = N_t \cdot \lambda \tag{12.1}$$

where N_t is the current population, N_{t+1} is the population in the following year, and λ is the growth rate parameter. A new population of 100 mussels last year increased to 145 individuals this year.

(a) What is the value of λ? *Hint:* Solve equation 12.1 for λ.
(b) Write a program that models the change in the population in this lake up to N_5 (note that $N_1 = 100$).
(c) Create a professional-looking graph of this population. Note that it is growing in discrete time.

5. We've spent time in this text thinking about p-values (you might want to revisit the definition for the p-value). Write a program that tests whether this works for the standard t-test. Below is a suggested algorithm to solve this problem.

(a) Create two samples of ten random numbers each, both drawn from the normal distribution. One should have $\bar{x}_2 = 0$, sd = 1; the other $\bar{x}_1 = 1$, sd = 1.
(b) Perform a t-test on your two samples.
(c) Store your t- and p-values in two separate arrays (e.g., my.t.value and my.p.value).
(d) Use a `for` loop to generate two new samples, perform a t-test, record the t-value, and count how many of the t-values are as extreme as or greater than your t-value from your original test. Do this test 1,000 times. Note that you do not want to write each number down! Have **R** record the number of times the new t-value is greater than your original t-value.
(e) Print to the console your original p-value and the proportion of times you got a t-value as great as or greater than the original t-value. Your proportion and the p-value from the first test should be similar.

6. The following equation is called the logistic map. This equation exhibits a wide range of extremely interesting dynamics.

$$N_{t+1} = \lambda \cdot N_t \cdot (1 - N_t) \tag{12.2}$$

When λ lies in the range $3.57 < \lambda < 4.0$, the series can produce chaotic dynamics. Create a graph of what happens when $\lambda = 1.5, 2, 3,$ *and* 3.75 over 100 time steps. Begin with $N_1 = 0.5$ for each simulation. Note that $0 \le N \le 1$.

7. Srinivasa Ramanujan, a famous Indian mathematician who died at the age of 32, proposed this estimate for π.

$$\pi = 1 / \left(\frac{2\sqrt{2}}{9801} \sum_{k=0}^{\infty} \frac{(4k)!(1103 + 26390k)}{(k!)^4 396^{4k}} \right)$$

(a) Determine the estimates for π for $k = 0 \rightarrow 5$. Note that you will need to use several levels of parentheses and make sure they're placed correctly!

(b) Graph your estimates of π as a function of k. Include a reference line at the value of π (`pi`).

CHAPTER THIRTEEN

Final Thoughts

I HOPE YOU'VE LEARNED A LOT! As we discussed, this science stuff is challenging. Perhaps you learned that **R** is challenging but manageable. You've learned how to install this program, and to install a front end (RStudio). You've learned how to extend this program with a few of the thousands of available packages. I hope you also have learned that you can solve problems using **R**. And I hope you've learned how to design better experiments and manage and analyze the data your get from those explorations.

Many people have gone through what you have. And most biologists today have encountered or used **R**. Many have said, "I'm not doing that" (only to quietly come on board). But you didn't say that—you've done it! You've had to look the beast in the eyes and come out victorious (at least sometimes, I hope). I commend you for your work. You should keep refining this skill. You can use **R** for simple calculations (what's the $\sqrt{5}$?). You can use it to make a quick graph of a function as well as make professional visualizations, solve differential equations, and solve just about anything quantitative you can think of. Put this skill on your resumé, and continue to solve problems.

In addition. I hope you've learned to be skeptical. Skepticism means you're not sure about something until you've seen the data. Good scientists are skeptical. Don't be cynical, which is the act of rejecting an idea because it exists. And I hope you've learned to be careful *believing* what people say. Instead, you're prepared to say, "Interesting, but I'd like to see the data for myself." So take those data, view them graphically, and test hypotheses yourself because you have access to and know how to use **R**, the statistical and programming choice of biologists worldwide.

13.1   Where Do I Go from Here?

I hope you're interested in learning more about how to solve problems in biological sciences. There are large numbers of biostatistics textbooks and books that introduce and extend various skills using **R**. Books on more advanced statistical techniques using **R** are coming out weekly. And consider taking a course (or another course) in statistics. But demand that the professor use **R**. You've come to realize that biological data analysis is not a spectator sport. And you should demand that you use this program in all your biology courses. Mathematics, statistics, and computer programming are modern tools for biologists.

Ultimately, there are many great sources of information on the topics covered in this book. You can find these in other books (see the Introduction) and from free online sources. If you have worked your way through this book and given the exercises the good old college try, then you're ready to handle the problems you'll find elsewhere. If the techniques you need are not in this book, you should be well versed in finding solutions to your problems. I often do a Google search by starting off with the letter "r" and then typing in some keywords. I rarely come up empty-handed with this approach.

I have not been able to introduce several topics in this book. An important frontier of modern biology has been collectively referred to as omics, which includes genomics, epigenomics, proteomics, and metagenomics, to name a few. One important tool for analysis of these systems is the package BioConductor (see http://www.bioconductor.org/), which runs within the **R** environment. The newer fields of big data, machine learning, and artificial intelligence are often tackled using **R**. You also may be interested in extending your programming skills with the Python language (see https://www.python.org/). Your work on developing how to solve these problems will serve you well in your career.

If you find the quantitative analysis of data and the building of models interesting and valuable, you're on your way to thinking like a modern biologist who seeks to better understand the complexity of biological systems. I recommend that you consider opportunities in the quantitative sciences. There are a variety of different opportunities to pursue as an undergraduate, including getting internships and research positions in what are called research experiences for undergraduates (REUs) (see https://www.nsf.gov/crssprgm/reu/).

One last point: achieving success requires hard work. No one starts answering challenging questions in science and says, "This is so easy!" Those who are, or have been, successful, however, might make it look easy. It's not! All of these successful people worked really hard. You too can do this. It takes focus and it takes time. But the payoffs will be many. You're on your way! Good luck!

APPENDIX

Solutions to Select Problems

Note that, for some of these answers, the code that generated them has not been shown.

Chapter 1

1. (a) ```
> sqrt(17)
```
```
[1] 4.123106
```

   (b) ```
> log(10,base = 8)
```
```
[1] 1.107309
```

   (c) ```
> exp(10)
```
```
[1] 22026.47
```

   (d) ```
> x = 3 # first, set x = 3, then use it
> 1/17 + (5*x + 7)^2 + log(17)
```
```
[1] 486.892
```

3. Using the provided values, the Earth has a volume of 1,083,207,266,253.2 km^3. There are too many significant figures, however, so we should just report the volume as 1.083207e+12 km^3.

5. The values should be
```
 [1]  1.00  1.41  2.00  2.83  4.00  5.66  8.00 11.31 16.00
[10] 22.63 32.00
```

8. a. The help says there are 7981 observations. Using the `length()` function, I got 7980. I don't know why it's wrong.

   b. ```
> plot(treering)
```

   c. The dominant period is 1,600 years.
```

Chapter 2

1b. When I do this, I get **84.87**.

2a. You should enter these data so they resemble those below.

	A
1	Mass
2	2.17
3	1.53
4	2.02
5	1.76
6	1.81
7	1.55
8	2.07
9	1.75
10	2.05
11	1.96

2f. When I do this, I get the following:

```
   Min. 1st Qu.  Median    Mean 3rd Qu.    Max.
  1.530   1.752   1.885   1.867   2.042   2.170
```

Chapter 3

1a. `> cheetahs = c(102, 107, 109, 101, 112)`

1c. `> order(cheetahs, decreasing = T)`

```
[1] 5 3 2 1 4
```

1e. `> signif(cheetahs,2)`

```
[1] 100 110 110 100 110
```

2e. `> head(mussels) # sorted dataframe after using the order()`
    `> function`

```
   Mass pH.trmt
15   87    High
12   83     Med
18   82    High
16   78    High
21   77    High
20   73    High
```

The left-most column shows the original rows for the mussels dataframe.

2h. `> subset(mussels, Mass > 80)`

```
     Mass pH.trmt
15    87    High
12    83    Med
18    82    High
```

Chapter 4

1a. The mean = 11.0 The standard deviation = 0.194. The median = 11.0.
SEM = 0.079. IQR = 0.25. The range = 0.5. CV = 1.76

1c. *Hint:* Summary statistics that assume the data are normally distributed rely on the mean as being a good measure of central tendency.

1d. They are normally distributed (W = 0.953, p = 0.76).

6. The mice have a CV = 31.9%.

Chapter 5

1a. These conversions can be looked up, but you'll have to use code to calculate these.

1b. Here's my version of the graph. Would it matter if I had switched "diam, ht" to "ht, diam"?

```
> plot(diam,ht, pch = 16, las = 1, cex.lab = 1.5,
+      xlab = "Diameter (cm)", ylab = "Height (m)")
```

3b. Here's the result for sample A:
`shapiro.test(M$A)`

```
> shapiro.test(M$A)
```

```
        Shapiro-Wilk normality test

data:  M$A
W = 0.74466, p-value = 0.02654
```

4.
```
> control = c(2,3,4,5,6,7)
> trmt = c(5,3,4,5,6,9)
> boxplot(control,trmt) # You'll need to improve this graph
```

Chapter 6

1b. That's a parameter because it's reporting the central tendency measure for the population (all students that took the test). It is not a sample of students that took the test. This, along with the variance estimate, are usually incorrectly presented as statistics.

2a. In general, we would consider $p = 0.06$ *not* to be statistically significant. However, if $\alpha \geq 0.06$, then $p = 0.06$ would be considered statistically significant. Notice that we need to use the "$\geq$" symbol and not "$>$."

6. The results might extend to females. This is where, as a student, you might consider conducting the same study on female rats or another mammal species. As the results stand, they extend only as far as the male rats tested. These might have been all members of the same line of rats and might not apply to other rat lines or rat species. It's also questionable to consider and report on beliefs in science.

Chapter 7

1a.
```
> moths = c(1916,1563,1436,6035,3833,5031,13326,3130,6020,1889)
> shapiro.test(moths)
```

```
        Shapiro-Wilk normality test

data:  moths
W = 0.78848, p-value = 0.0105
```

```
> shapiro.test(log(moths))
```

```
        Shapiro-Wilk normality test

data:  log(moths)
W = 0.93315, p-value = 0.4796
```

This suggests these data are not normally distributed but they are after log-transforming them. Therefore, we can proceed with a parametric test.

1b. Here are our hypotheses:

H_0: $\overline{moths} \geq \overline{new}$
H_A: $\overline{moths} < \overline{new}$

1c. This is a one-tailed test for whether the sample is *less than* 10,000. Note that I've used a t-test on the logged data *and* that I've log-transformed the test value.

```
> t.test(log(moths), mu = log(10000), alt = "l")
```

```
        One Sample t-test

data:  log(moths)
t = -4.652, df = 9, p-value = 0.0005993
alternative hypothesis: true mean is less than 9.21034
95 percent confidence interval:
    -Inf 8.565095
```

```
sample estimates:
mean of x
 8.145499
```

1d. The new number of moths is significantly higher than previous samples (t = −4.652, df = 9, p < 0.001).

3a. I chose to create a new variable of the differences (lf.diff). This sample is normally distributed (p = 0.696).

Chapter 8

1c.
```
> milk.dat.stacked = stack(milk.dat)
> names(milk.dat.stacked) = c("Bac.count","Farm")
> head(milk.dat.stacked)
```

```
  Bac.count    Farm
1        24  Farm.1
2        15  Farm.1
3        21  Farm.1
4        27  Farm.1
5        33  Farm.1
6        23  Farm.1
```

1d. Here's the code to accomplish this. The results are not shown.

```
> tapply(milk.dat.stacked$Bac.count, milk.dat.stacked$Farm,
+        shapiro.test)
```

2b. Here's how to subset the dataframe in a way that collects just the fish length values for the cold temperature and high pH treatment data into an array.

```
> a = subset(my.sticks, Temp == "C" & pH == "H")$Length
> shapiro.test(a)$p.value # repeat for other samples
```

```
[1] 0.492481
```

2c. Here's the graph for just the main effect of temperature on stickleback lengths.

```
> library(plotrix)
> M = tapply(my.sticks$Length,my.sticks$Temp, mean)
> SD = tapply(my.sticks$Length, my.sticks$Temp, sd)
> SEM = SD/sqrt(6) # there are 8 observations per sample
> CI95 = qt(0.975,5)*SEM
> a = barplot(M,xlab = "Temperature",ylab = "Length (cm)",
> las = 1,
+     ylim = c(0,6), names = c("Cold","Warm"),cex.lab = 1.5)
> abline(h=0)
> plotCI(a, M, CI95, pch = NA, add = T)
```

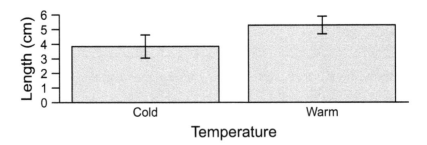

Chapter 9

2b.
```
> femur = c(38,56,59,64,74)
> humerus = c(41,63,70,72,84)
> plot(femur,humerus)
> cor.test(femur,humerus)
```

```
            Pearson's product-moment correlation

data:  femur and humerus
t = 15.941, df = 3, p-value = 0.0005368
alternative hypothesis: true correlation is not equal to 0
95 percent confidence interval:
 0.910380 0.999633
sample estimates:
      cor
0.9941486
```

The femur and humerus lengths for these fossils follow a strong, positive correlation (r = 0.994), which is highly significant (t = 15.9, df = 3, p < 0.001).

2e. This is consistent with these fossils being from the same species.

4a. You should first check out the different seed sources available in the dataset:

```
> unique(Loblolly$Seed) # this asks for the unique types
```

```
 [1] 301 303 305 307 309 311 315 319 321 323 325 327 329 331
14 Levels: 329 < 327 < 325 < 307 < 331 < 311 < ... < 305
```

Now we can subset the dataframe. I'll choose to do this for seeds from source 321:

```
> Loblolly.321 = subset(Loblolly, Seed == 321)
```

Chapter 10

1a.
```
> obs = c(295,86) # observed counts
> expP = c(0.75,0.25) # expected probabilities
```

2a. You should find that $\chi^2 = 0.47$.

3a. You should find that $\chi^2 = 36.9$.

Chapter 11

1.
```
> my.stats = function(x) {
+     my.mean = mean(x)
+     my.sd = sd(x)
+     my.SEM = my.sd/sqrt(length(x))
+     return(data.frame(my.mean,my.sd,my.SEM))
+ }
```

3. Below are some useful lines of code:
```
> mpg = c(44,52,58,63,58)
> speed = c(20,30,40,50,60)
> plot(speed,mpg)
> mod = lm(mpg ~ poly(speed,2))
> newx = seq(20,60,by = .1)
> lines(newx,predict(mod, list(speed = newx)),lwd = 2)
```

4a.
```
> N0= 50
> r = 0.82
> K = 1000
> curve((K * N0 * exp(r*x))/(K + N0 * (exp(r*x)-1)),0,10,
+       ylab = "N", xlab = "Time", cex.lab = 1.5)
```

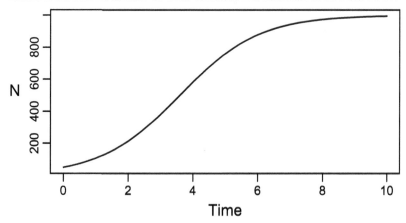

5a. There are at least two ways to subset the data. Below I show these for
the untreated sample (`untr`).

```
> attach(Puromycin)
> untr = Puromycin[state == "untreated",] # or use
> # dat = subset(Puromycin,state == "untreated")
> tr = Puromycin[state == "treated",]
> par(mfrow = c(1,2))
> plot(tr$conc,tr$rate, ylim = c(0,220), pch = 16,
+       xlab = "Concentration", ylab = "Rate",
+       main = "Treated")
> plot(untr$conc,untr$rate, ylim = c(0,220), pch = 16,
+       xlab = "Concentration", ylab = "Rate",
+       main = "Untreated")
> par(mfrow = c(1,1))
```

Chapter 12

2. This can be done easily without a `for` loop:

```
> sum(1:100)
```

```
[1] 5050
```

This is the correct answer, but it's cheating! To accomplish this the *hard*
way, using a `for` loop, we can do the following:

```
> the.sum = 0 # recall "sum" is a function, so avoid
> for (i in 1:100) {
+    the.sum = the.sum + i
+ }
> cat("The sum is ",the.sum,"\n")
```

```
The sum is  5050
```

4b. Here's my approach to modeling the growth of the zerbra mussel population:

```
> lambda = 145/100
> N = numeric(5)
> N[1] = 100
> for (i in 2:5) {
+    N[i] = lambda * N[i-1]
+ }
> N
```

```
[1]    100.0 145.0 210.2 304.9 442.1
```

6. Here's the graph of the logistic map when $N_1 = 0.5$ and $\lambda = 3.75$.

```
> N = numeric(100)
> N[1] = 0.5
> lambda = 3.75
> for (i in 2:100) {
+    N[i] = N[i-1]*lambda*(1 - N[i-1])
+ }
> plot(N,type = "l", xlab = "Time Step", ylim = c(0,1),
+        cex.lab = 1.5, las - 1)
```

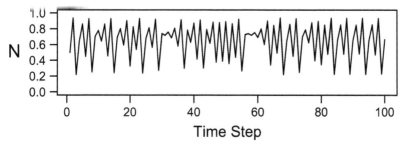

Bibliography

Adler, J., 2012. *R in a nutshell: A desktop quick reference.* O'Reilly.

Crawley, M., 2012. *The R book.* Wiley.

Fagoonee, I., H. B. Wilson, M. P. Hassell, and J. R. Turner, 1999. The dynamics of zooxanthellae populations: A long-term study in the field. Science, **283**:843–845.

Focht, D. R., C. Spicer, and M. P. Fairchok, 2002. The efficacy of duct tape vs. cryotherapy in the treatment of *Verruca vulgaris* (the common wart). Archives of Pediatrics & Adolescent Medicine, **156**:971–974.

Gotelli, N. J., and A. M. Ellison, 2012. *A primer of ecological statistics*, 2nd ed. Sinauer Associates.

Grolemund, G., 2014. *Hands-on programming with R: Write your own functions and simulations.* O'Reilly Media.

Hartl, D. L., and D. J. Fairbanks, 2007. Mud sticks: On the alleged falsification of Mendel's data. *Genetics*, **175**:975–979.

Hartvigsen, G., 2011. Using R to build and assess network models in biology. *Mathematical Modeling of Natural Phenomena* **6**:61–75.

Houck, M. A., J. A. Gauthier, and R. E. Strauss, 1990. Allometric scaling in the earliest *Archaeopteryx lithographica*. *Science* **247**:195–198.

Hurlbert, S. H., 1984. Pseudoreplication and the design of ecological field experiments. Ecological Monographs, **54**:187–211.

Marshall, W. F., H. Qin, M. R. Brenni, and J. L. Rosenbaum, 2005. Flagellar length control system: Testing a simple model based on intraflagellar transport and turnover. Molecular Biology of the Cell, **16**:270–278.

Matloff, R., 2011. *The art of R programming.* No Starch Press.

Mendel, G., 1866. Versuche über pflanzen-hybriden. Verhandlungen des naturforschenden vereines. Abh. Brünn, **4**:3–47.

Meys, J., and A. de Vries, 2015. *R for Dummies*, 2nd ed. For Dummies.

Nelson, W. A., O. N. Bjornstad, and T. Yamanaka, 2013. Data from: Recurrent insect outbreaks caused by temperature-driven changes in system stability. Dryad Digital Repository, **341**:796–799.

R Core Team, 2020. *R: A language and environment for statistical computing*. R Foundation for Statistical Computing, Vienna, Austria.

Rosner, B., 2015. *Fundamentals of biostatistics*. Cengage Learning.

Silver, N., 2012. *The signal and the noise: Why so many predictions fail—but some don't*. Penguin Press.

Templeton, C. N., E. Greene, and K. Davis, 2005. Allometry of alarm calls: Black-capped chickadees encode information about predator size. Science, **308**:1934–1937.

Triola, M. M., M. F. Triola, and J. Roy, 2017. *Biostatistics for the biological and health sciences*, 2nd ed. Pearson.

Venables, W. N., D. M. Smith, and the R Core Team, 2018. *An introduction to R*. Network Theory Ltd.

White, C. R., and R. S. Seymour, 2003. Mammalian basal metabolic rate is proportional to body mass$^{2/3}$. *Proceedings of the National Academy of Sciences* **100**:4046–4049.

Zar, J. H., 2009. *Biostatistical analysis*. Prentice Hall.

Zhang, X., Lei, B., Yuan, Y. et al. Brain control of humoral immune responses amenable to behavioural modulation. *Nature* 581, 204-208 (2020). https://doi.org/10.1038/s41586-020-2235-7

Zuur, A., E. N. Ieno, and E. Meesters, 2009. *A beginner's guide to R*. Springer.

Index

Printed in the USA
CPSIA information can be obtained
at www.ICGtesting.com
JSHW051458221024
72172JS00011B/100